4.18.79

Stealing the sun is too ridiculous to imagine—or is it?

Monopolization of the sun. A virtual enslavement of generations to come...control of a man's life through the control of the energy necessary to his existence.

Giant corporations are setting out to...own the sun outright. The solar market equals 71 times Exxon's annual sales.

You and your grandchildren will end up paying for the sunshine that falls on your yards!

The suspicion is almost unavoidable that giant firms, because of their large investments in nuclear technology hope that solar energy will not gain rapidly—Senator Gaylord Nelson of Wisconsin.

Today, you can buy a solar furnace for $3000 to $6000 yet the government wants to install 650 solar heating systems on homes for a cost of $76,923.00 *each* as a test.

The problem with the fuel shortage is the way we've been lulled into disbelief. We always talk about it in the third person...somewhat like death.

The big lie: "Solar equipment is costly. Only through leasing can the ordinary man afford it." If you believe that...you've been suckered again.

If oil companies get depletion allowances, the homeowner deserves an allowance for *not* depleting.

1

THE SOLAR CONSPIRACY

THE $3,000,000,000,000 game plan of the energy barons' shadow government

by
John H. Keyes

MORGAN & MORGAN, Publishers
Dobbs Ferry, New York

049

/9 10650 9

International Standard Book Number 87100-095-4
Library of Congress Catalog Card Number 75-24833

Copyright 1975 by John H. Keyes
Published by MORGAN & MORGAN, INC., Publishers
145 Palisade Street, Dobbs Ferry, New York
Printed in U.S.A.

DISCLAIMER AND ACKNOWLEDGMENTS

Although I am the bearer of the slightly pompous title of Chairman of the Board of International Solarthermics Corporation of Nederland, Colorado, this book was a private and personal undertaking. The views contained herein are my own and do not represent the views of any other person, corporation or political or private entity. Any of these views which coincide with the views of other persons, corporations or etc., are purely coincidental and represent a very low statistical probability, no doubt. This is my work and no one else's.

I am tremendously grateful for the assiduous red pen manned by Jon Hanson and for the loving support and patience of Erika, Betty, Heidi-Rika and John-Michael.

2058002

FOREWORD

I am an average, apathetic, middle-class American, one of the "scuttlefish" referred to by Harlan Ellison in the *Glass Teat*.

Watergate was a shocker, and I alibied for President Nixon long past logic, because I have an inherent faith in our system which borders on blind, unthinking loyalty to the country and the leadership of the country. Typical. I *want* to believe in the integrity of our President, Senators and members of the House of Representatives. *You* may jeer at that, and justifiably so.

I wish that I could rationalize that I am a product of an educational system which molded me into that passive-acceptance, authority-respecting syndrome. But that would be pure nonsense.

The Dewey-eyed social workers who would excuse *any* action of mine as someone else's fault—my parents', my schools', my society's—make me impatient. What I opt to do is *my* responsibility. I can't buy the all-encompassing generality that a "man is the product of his environment". Perhaps true in its more trivial implications: "I was raised in the Midwest, so I speak with a Midwestern twang." But I feel very strongly that *a man takes only that which he chooses to from his environment*, rejecting daily huge quantities of input from that environment as fallacious.

A more valid analysis might be that the environment is the product of the people and natural forces in it.

For the past two years I have been involved twenty-hours-a-day, seven-days-a-week in solar energy research at a privately funded research and development facility called International Solarthermics Corporation. We invented and developed the backyard solar furnace.

The word "naivete" never realized its full potential until it was applied to that small group of researchers involved in the final testing of that solar furnace, the making of it a production-ready item, and the licensing under patents pending of many independent manufacturers in competition with each other to make and market this exciting new breakthrough product.

Instead of being greeted with "huzzahs", the backyard solar furnace seemed to act like a red flag stimulus designed to prompt the anger of the people already working in the field of solar energy research. No hero's welcome. Well, just sour grapes said I.

The solar furnace was attacked by learned PhD's from the date that marginal press attended its development. And although these attacks were clothed in engineering language, they were incredible from an engineering standpoint. Many times, the attacks were addressed to laymen in impressive-sounding, but nonpertinent, if not invalid, scientific language from a respectable source.

If one saw a professional house painter painting a house white, and suddenly attacked the project, saying, "You aren't painting the house white!", the ludicrousness of the attack is such that the poor house painter is left with but one retort. "Oh, yes I am!" One can imagine the house painter appealing to several bystanders, asking, "It *is* white, isn't it?"

Indeed, some sour grapes comments *were* involved. But could other motives have been involved? It is these *other* motives for attack upon the backyard solar furnace to which this book will hopefully call attention.

But first an examination of the sour grapes. In retrospect it is easy to see why such responses were elicited. The backyard solar furnace was *ten times smaller* than other systems developed to that date. Respected scientists had been wrestling with the problems of harnessing the sun for years, and if they could not build a practical system with less than *1,000 square feet of solar collector*, how could some researchers—in a small mountain community headed by a man with (horror of unscientific horrors) a mere Bachelor's degree in philosophy—have possibly developed a practical system with *less than 100 square feet of collector?*

It *was* a slap in the face to the academic community, a label on their past efforts as incompetent.

The attacks continue to this date without respite. Fortunately many attackers suspended final judgment until they had visited our research facility and examined the data and data collection methods being utilized. Each of them who has taken the time to investigate carefully has become a "convert" and an advocate for the system.

The attacks, however, are unimportant to this book, and no defense of the backyard furnace will be presented in the pages which follow. The incidents are mentioned here to chronicle the onset of paranoia which followed. Sour grapes were to be expected, and with time and communication, they would slowly fade away. Our company held a special meeting over a year and a half ago and decided not to counterattack the baiters with libel and slander suits. The tone of the attacks showed them to be largely emotional, and such suits would only serve to embarrass these men for whom we had a great deal of respect, due to their loyalty to and efforts expended for the cause of solar energy utilization. We have to this date adhered to that early decision. At times with difficulty, for our integrity was impugned on many occasions, and it is difficult not to meet such charges with emotional countercharges. We bit our tongues.

In this continuing atmosphere of ceaseless attacks—which was beneficial in one way, because we had to be twice as cautious with our work—the furnace was refined, tested and then subjected to a series of final verification tests which were completed in September 1974. It took nearly five months after that to reduce and analyze all of the data gathered during the previous year. This was finally accomplished, however, and published in a proprietary package of technology transfer. The first of many manufacturers was licensed to produce under our letters of patent in February 1975.

It was at about this time that the character of the attacks changed significantly. Up until this time, the attacks had been directly delivered. That approach, for the most part, stopped. Now we heard about attacks, but in the third person, always.

Someone with whom we had talked would "hear" from a "respected source" that our system violated laws of physics. (The ridiculousness of such a charge may have to be explained to the layman. The laws of physics are inviolable. The proper scientific attitude is that "violation" of these laws is a *logical possibility*—but we have never yet seen *any* process which violates them. The real world attitude is that they will *never* have exceptions—indeed, this is reflected in patent law, since no device that violates these laws of physics is patentable. Perpetual motion devices, for example, may not be patented.)

Obviously, the backyard solar furnace does not violate *any* laws of physics. In the five patents pending on the device some thirty claims have already been approved, so the patent office has missed violations of the laws of physics if they are there!

The layman doesn't know how to interpret these "backdoor" attacks, however. Mr. Joseph Dawson, assistant to Virginia Knauer, the President's advisor on Consumer Affairs, for example, related one such third party attack to me. He refused to disclose the source, but stated that it was a respected scientist. To Mr. Dawson's benefit, he did not accept the charge out of hand, but did ask me about it. I could only answer that the charge was ludicrous and untrue. I offered to pay for a registered professional engineer of his choice to check out the charges. He declined.

This new vector for the attacks, however, was not noticed immediately. We were resigned to the attacks, and our sensibilities, admittedly, were blunted.

During the period from September to date, we experienced a series of break-ins at our research and development facility. *But nothing was ever stolen!* Occasionally we were able to determine that papers on desks had been shuffled, but no other signs of entry could be found. The burglar alarm would go off, and occasionally we would find a locked door standing open. At first, we, and the local town marshal, thought that the alarms were ascribable to malfunctions of the equipment. Then, on one occasion of routine maintenance on the alarm system, the repairman found the circuitry on the alarm in one building had been rewired. On another occasion, a neighbor to the complex noted a helicopter hovering fifty feet above the ground behind the research complex. On another occasion an *unmarked* fixed-wing airplane flew a half-dozen low-level passes over the complex, wing-over on each pass, before disappearing. Finally, after an alarm in January, where the handle of a locked fire-file was found broken, again with nothing stolen, we hired armed guards to patrol the premises during the times that staff was not around. Since that step was taken, there have only been two alarms. Significantly, they both occurred at times of shift change on weekends, between the time that the guards went off duty and the day staff arrived. Again, nothing stolen. *If* espionage is going on, who is responsible? And why would a small mountain research firm be a target? I don't even have a good guess as to the answer to the former question. The latter might be attributable to our research into solar electrical generation and solar automobile engines.

During the past two years, jokes have been made about solar energy. "They'll want to charge for the use of sunshine", or, "The next thing you know they'll be wanting to tax sunshine." But no seriousness was attached to such statements. One would have to be paranoid in the extreme to conjure up a serious conspiracy to monopolize sunshine!

Although many news releases concerning public utilities indicated their strong desire to get involved in delivery of equipment, the immensity of such an attempt simply made it unthinkable.

It was *one* solar energy conference that opened my eyes. Finally the issues and the people and corporations involved became clear.

That conference was "Solar Expo 75" held at the Sheraton Park Hotel May 27-30, 1975, in Washington, D.C. It was ostensibly a trade show, the first annual exposition of the

Solar Energy Industries Association (SEIA), a trade association organized about a year ago by a Washington PR firm headed by James Ince, co-hosted by the Energy Research and Development Administration (ERDA) and the Federal Energy Administration (FEA).

The players were in attendance in force at the conference. As I met with some of them, and other staff members and officials on the "hill" in Washington, D.C., a "picture" emerged concerning the "game" being played.

This book is an attempt to share that "picture" with you. Describing the various players and their interests where the latter are attributable. Hopefully to integrate the complex of their motives and goals as they interrelate with each other, and to integrate the complex of arenas which have sprung up in response to the impending energy shortage.

I have named the players. I take responsibility for that, despite some advice to the contrary, for to do less, in my opinion, would be a cop out.

I am not an investigative reporter. I have neither the resources nor the staff to be one, let alone the inclination or talent. Therefore, in places, the conclusions I have arrived at may be without substantive documentation. In those places, I have relied upon analysis and logic, supplying horse sense, farmer-logic assumptions and premises concerning motives, relying on your objective assessment of such supplied premises to develop the picture.

I hope that one of the new generation of incisive investigative reporters who have grown as a class in the past ten years will take the ball and run with it from here. An in-depth substantiated and documented exposé is screamed for here.

If this book serves as that necessary stimulus to trigger such an investigation, it will have been successful. If it makes you angry—either at me or with me—it will have been partially successful.

John H. Keyes
June 15, 1975
Nederland, Colorado

INTRODUCTION

Imagine a science fiction writer sitting down at his typewriter and pounding out a story in which evil and malevolent bug-eyed monsters conspire maliciously to steal the sun from the solar system with a mysterious, powerful "tractor beam". Such a story would be destined for the pulp magazines at best, since the precept is unbelievable. Stealing the sun is too ridiculous to imagine—or is it?

The free sunshine falling on the Earth *is* being stolen from you. *Right now. Today.* Not with tractor beams, but with politics, misdirection, contradictions, confusion, lies, and legislation seemingly unrelated to sunshine. And much more easily than one would suspect.

This book is an exposé of what I feel to be a criminal conspiracy; a conspiracy to steal the sun from you. A giant undertaking, to be sure. But then it is the corporate giants of this world who are engineering the theft.

Obviously even poor, underrated Mr. and Mrs. Consumer can be eventually riled up and become dangerous, as many tyrants and opportunists in the past have learned too late, to their misfortune.

Therefore, no overt theft, this, but a subtle and devious plan with all of the misdirection and obfuscation of which the giant corporations are capable.

It would be nice if I could expose that planned theft simplistically in all of its pustulance. But the real world is not like that. The issues are complex and intertwined. Other players are inadvertently operating in that arena, confusing the issues, becoming unknowing advocates for a policy they would abhor if they realized the consequences. Solar energy, unlike many other technological advances of the past, really portends something new and different for the future. Its development means *personal* independence for the average, ordinary man. *That* is revolutionary in its import!

We have all witnessed, over the past decade, the ease with which captive scientists, owned by corporate giants, prominently displaying their PhD's as badges of omnipotence and omniscience, have loudly disclaimed the hazards of widespread proliferation of nuclear power plants, have argued for catalytic converters, poo-pooed the hazards of just that product which their employer happens to manufacture or sell, and in various ways served the private interests of their employers to the absolute detriment of the general public.

The morality of the multinational energy suppliers is often misinterpreted. Their goal is not just *profit*, as some accusers espouse. Rather, the goal is *power*—over people and over nations. In fact, in many significant areas they have already gained and hold significant political power. How much of the political business conducted in Washington happens to work out to the benefit of the giant corporations?

Who wields the power of government in America? The White House and Congress, right?

On my last trip to that city, a veteran Washingtonian laughed at this and said, "The legislators come and go every two years—their 'power' is illusory. The real power lies in the staff members on the hill. They stay here for thirty, forty, fifty years. And it is in their hands that the government rests. They are supposedly the advisors, the counselors, but in actuality they draft the legislation, hold the hearings, conduct the investigations, sort the mail and exercise a near total control over the legislators."

At the start of that last visit, I still disbelieved that cynical assessment. By the end of the visit, I knew it to be valid in my bones—not from actual evidence, but from veiled threats, contained not in words, but in their delivery. This was a powerful man—don't alienate him even if you have never heard of him—and it communicated itself by the assured way he talked to you, the way others talked to him, the way "important people" gravitated to his presence. . . .

It was a trauma, this recognition. It re-emphasized the farmer's belief that honor and politics are mutually exclusive categories. That the politician who is the most sincerely honest and honorable is the one most subject to being callously manipulated, the most damningly ignorant and naive.

The positive occurred as well. The conspiracy which had been repeatedly laughed off unveiled itself—as the goals and motives of the parties became clear.

Monopolization of the sun, of sunshine. A virtual enslavement of generations to come. An enslavement no less perverse simply because it did not represent direct corporeal bondage, but the more insidious control of a man's life through the control of the energy necessary to his existence. The grasp for power is no less evil because it does not aim at occupying the throne, but much more deviously controls the throne, the body legislative, the people as a whole and operates unseen, unrestricted and unchecked behind the scenes.

Dramatic language perhaps. Don't expect, however, passive acquiescence from this citizen while he, his family, his friends and his descendants are being gang-banged by the giant corporations and their willing cohorts in the seat of national government!

The distribution of this book will be widespread, but with a feeling of helplessness. One begins to wonder if the public apathy so widely disclaimed is not being deliberately fostered by a giant corporation-controlled television campaign of the dead-boring designed to make the most trivial, the most inept, the most passé, the most cliché, the sublime in our society. (It *is* said that the average American watches television over six hours *every* day.)

The degradation of personal liberty continues in these passive Seventies. Even the frightening unrest of the Sixties is to be preferred to the unmitigated and tranquil *acceptance* of Watergate and its concomitant corruption with an apparent and accepting bovine smile, "That's the way politics are."

Friend, wake up. They're getting ready to *really* rip you off this time. *Wake up.*

XV

CONTENTS

CHAPTER 1 1
The Stakes of the Game 3
CHAPTER 2 9
Putting the Mystery into Solar Applications 11
CHAPTER 3 17
Servants of the People 19
CHAPTER 4 25
Energy Blackmail to Force Implementation of Nuclear Power ... 27
CHAPTER 5 35
Engineering Humbuggery 37
CHAPTER 6 45
David and Goliath 47
CHAPTER 7 59
The Contradictions—Why? 61
CHAPTER 8 65
Is There a Fuel Shortage? 67
CHAPTER 9 75
The Question of Retrofit 77
CHAPTER 10 81
Consumer Protection and "Certification" 83
CHAPTER 11 91
Monopolization of the Sun 93
CHAPTER 12 99
The Leasing Lie 101
CHAPTER 13 107
Regurgitation 109
CHAPTER 14 113
Building A Battle Plan 115
CHAPTER 15 121
The Solar Conspiracy 123
APPENDICES
A. 17 Ways to Cut Your Fuel Bill in Half A1-A12
B. Advice to the Builder or Homeowner
   Considering Solar Heating B1-B11
C. Advice to the Architect Considering Solar Heat C1-C11
D. Myths and Facts About Solar Heating D1-D10

CHAPTER ONE

"Why do they follow us, old man?" The boy's dirty face looked puzzled in the flickering light from the small fire.

The old man chuckled. "Not for the reasons you would think, boy. Not for the reasons they think. And maybe not for reasons at all."

He sat hunched against the chill mountain wind sighing quietly through the darkness around them, oblivious to the icy points of light suspended in the obsidian above them.

Coffee bubbled in the rusted can nestled in the fire.

The gnarled tree which served as a rustic chair formed a faint mosaic of reds and oranges behind him.

The boy huddled still, his back against a boulder across the fire.

"The sun will come soon, boy, and that's ours. It is our strength and power. And maybe that is why they follow —to wrench that power from us."

1

THE STAKES
OF THE GAME

If one is going to suspect a conspiracy, one must understand the motives. The giant corporations are not pikers, so the table stakes have to be huge. We will examine those stakes: They are immense—the biggest ever played for since scientists played Russian roulette with the first atomic detonation, being "pretty sure" that the explosion would not set off a chain reaction which would destroy the world.

The motive: huge stakes

Ralph Nader recently suggested that if the government would just deed the sun to Exxon, practical solar energy utilization would be immediate. The point was absolutely correct. There are **no** technological impediments to the introduction of solar utilization **today** ...The sardonic humor of the statement is lost, however, when one sees that giant corporations are setting out to do just that: own the sun outright. Who needs a deed if he has all the production?

Giant corporations out to own the sun

And "all of the production" is a tremendous pot in the middle of the card table. Depending upon whose statistics you decide to believe, there are from 65 million to 80 million existing single family homes in the United States. Supplying only the solar heating **equipment** for those homes would be a $325 billion to $1 trillion business!

$3 trillion market potential

3

When one includes the equipment necessary for our schools, public buildings, apartment houses, warehouses and factories, the sales volume possibility exceeds several trillion dollars. New construction alone would account for an additional 20 to 50 billion dollars, **annually**.

Market equals 71 times Exxon's annual sales

To put that in perspective, Exxon's gross sales last year were $42 billion, General Motor's $31.5 billion, Ford's $24 billion, Texaco's $23 billion, and Mobil's $19 billion. Those are the big 5. If they, for example, geared up to install solar equipment, and between them shared the total market, and installed the equipment over the next ten years, it would **double** their combined sales annually for that full ten year period.

Huge profit potential not enough

One would think that the huge profit potentials involved in that volume of sales would assuage the appetite of the most greedy, profit-hungry corporation. Not so.

Lease solar equipment instead

Contemplate instead: don't **sell** the equipment. **Lease** it instead. Why? Then, even with a modest 10% return via lease revenues, one can project a 100-300 billion dollar annual income, with a two per cent or so increase each year due to new construction.

Better yet, —rent solar energy like the telephone

Every corporation in the United States envies the International Telephone and Telegraph philosophy. Don't even lease the equipment—**rent** it. Then you can up the revenues and still own the equipment **in perpetuity**. That's much better than leasing. With leasing you might have to include a purchase option at the end of the lease period.

4

Notice that it doesn't sound terrible: you'll be able to rent your solar equipment just like you rent the phone. The horror in that statement is well hidden: you really end up paying, forever and forever, and your children end up paying, forever and forever, and your grandchildren end up paying, forever and forever, **for the sunshine that falls on your yard!**

This concept, if you begin to see the signposts pointing to it in the stories you've read on solar energy, demonstrates that **not only are massive profits sought**—the stakes include **control and the power which is concomitant to control.**

Interestingly, the stakes get sweeter as we go along. For the government has already indicated a willingness (in principle at least) to subsidize the capital expenditures necessary to obtain the equipment. Numerous bills have been introduced in both houses to permit tax credits, tax deductions, tax exemptions on the installation of solar equipment.

Under the rules of the ultimate game, the taxpayer would pay, indirectly, through taxation for the equipment subsidies given to the owner of the equipment. The utility would own the equipment without having to pay for it. Then the taxpayer would supposedly bend over accommodatingly and rent the equipment **he** paid for back from the utility. Not just paying for the sunshine itself, but supplying an infinite percentage of profit to the utility. Walter Mitty never dreamed such an ambitious dream!

All future generations must pay for use of sunshine

Not just profit, but power

Government willing to subsidize solar energy

Taxpayer pays for equipment utility owns

No solar energy until after nuclear plants established

The issues, however, are complex. Solar monopolization **is** being sought, but within a specific time frame. In the meantime an even more hideous game is being played by the nuclear advocates. Since tremendous monetary commitments have **already** been made by the nuclear club members to the development of atomic energy utilization, they must **first** get that very specialized, uni-use equipment **installed, just to recover the massive investments already made in that field.** Then — and only then — can the solar monopolization game begin.

Develop fuel "emergency" to force nuclear plants

How do they hope to **force** the introduction of nuclear plants all over the country? It is the author's opinion that the scenario is relatively clear. Downplay, contradict, obfuscate and suppress the information available about the severity of the present fuel crisis. Allow the fuel shortage to develop into a full-blown emergency. **Then** the American public will **beg** to have nuclear plants installed anywhere — even near their homes.

Discredit gad-flies who attack nuclear implementation

This is a game of subtlety. A game of discrediting those who speak out. One can see, if he but looks, an active campaign aimed at discrediting Ralph Nader who had the gall to worry about the public's safety, protection of the plants from terrorists and the problems of waste disposal stretching over periods of time far in excess of recorded history to date.

Undermine the man's integrity

A campaign is being waged which is national in scope **and is working**. A whispering campaign which attacks Ralph Nader in **other** areas — "The Corvair wasn't **really** unsafe. He just attacked it to get publicity." Such a campaign plays on the envy of the famous, and is readily passed on by the average man, without investigation or thought.

News, or psychological strategems?

The same discrediting tactics are being used in the burgeoning new solar furnace industry. A solar-electric research firm in Southern California recently underwent a scathing attack, complete with indictments of the officers, with tremendous amounts of media coverage. After the courts had found the officers innocent, the announcements of that fact were small by comparison to the earlier trumpeting of charges. Deliberate? Perhaps not. We, as a race, enjoy the gossip of charges of evil, and are somewhat disappointed when the charges are discovered to be unfounded.

Another issue: the "fast buck boys"

Another issue is raised to really cloud the battleground. All over the country, the "blue suede shoe boys" are moving into the business of solar energy. The "fast buck" is scented by these artists, and they see the opportunity for tremendously big profits in this new industry.

How to recognize the honest man?

Unfortunately, these money promoters are not easily distinguished by the public from the honest but sometimes flamboyant entrepreneur investing huge sums of money and time in the frustrating business of pioneering a new industry. Consumer protection is needed. But the consumer is best protected by **educating** him, not by having governmental agencies come in to engage in "protecting" him **after** he has lost his money. Such agencies are always late.

Nuclear ploy: "consumer protection" to slow solar implementation

As we will see, however, one method of hindering implementation of solar energy utilization until after the nuclear gambit is completed is to trumpet "consumer protection" as a way of creating needlessly complicated "certification" procedures which will slow the honest manufacturer.

CHAPTER TWO

Their clothes were wet and tight as they struggled up the boulder-laden slope, across the slippery, treacherous talus. Upward was hard, thought the boy.

"Can we rest a moment, old man?"

"No, boy. Not yet."

The trees thinned and disappeared and it was cold.

The sun disappeared abruptly and the tumult of a mountain rain, fierce and unremitting, assailed them for daring to enter the timberline country.

Finally they found refuge in an ancient prospect hole carved by some optimist into the cold soil, vomiting a yellow refuse from its gaping maw. Lightning brightened the gray around them in a sporadic yet natural rhythm.

"Fire, old man?"

"No, boy. The sun will return in a few minutes." They wrapped their coats around their knees and waited.

PUTTING THE MYSTERY INTO SOLAR APPLICATIONS

Another interest group is making itself felt in the new solar arena, as well.

The governmental grant seeker.

Another player: the government grant seeker

Whether corporate or academic, the grant seekers must have a special grant proposal writing staff. It is a fact of life that the approved grant will be composed of, in equal parts, some research and some publicity. The committee issuing the grant will be impressed first by the professionality of the grant proposal itself but more importantly will see the reflected glory that the grant will bring to the issuer. Each project then must be a "first" in some way which will guarantee headlines.

Solar grants to similar residential projects

Despite the unwritten requirement for uniqueness, a large number of virtually identical single family home solar heating and/or cooling projects have been funded all over the United States over the past two years. Each used a hydronic collector—a collector using water running over or through the plates of the device—mounted on the rooftop, and each had a behemoth storage tank of water in the basement. Although such systems had been deemed uneconomically feasible for decades, they were being built again. (Presumably to show that they were **still** economically unfeasible.) Yet the publicity that attended each of these heralded

11

the "firstness" of it.

It is the nature of the grant business that public relations and press releases on the grant are necessary as a foundation to the obtaining of the **next** grant. Most of the projects, unfortunately, have more press than results.

The media make a contribution to the mysteriousness of solar heating inadvertently. The editor or news director wants something with headline appeal. Due to this necessary approach, the "mystery" of a solar system is heightened, in that it is very easy to shroud the device's practicality in difficult-to-understand technological language. The grotesque or unusual is, of course, the focus of the story.

This **approach** is furthered appreciably when the solar device itself **is** loathsome. What average American wants 1500 square feet of glass and aluminum on top of his 1500 square foot home? Or his basement filled from wall-to-wall and from floor-to-ceiling with gargantuan plumbing, valves, whirring pumps, etc? The picture conveys an ominous and hazardous air — a kind of Frankenstein's workshop atmosphere — to the solar heating equipment. What person in his right mind would want that thing crouching malignantly in **his** basement?

The publicity attendant upon the grants is perverting in another way. It insidiously engenders distortion from the participants as well. The City of Colorado Springs, Colorado, for example, with participation from a number of businessmen and corporations, built a "solar heated house", the

Phoenix Project, last year, completing it in May of 1974. The project spokesmen, probably unaccustomed to the herald and fanfare that accompany being in the public eye, got carried away.

A report to the National Science Foundation from James D. Phillips, Director of the Department of Public Utilities, dated 10 January 1975 was oddly at variance with published reports issuing from the Phoenix house group. The report states that the cost of materials was only $11,122.40 but a careful scrutiny shows that some materials must have been contributed in addition to this. For example, the list only contains one-fourth of the amount of glass required to glaze the collector.

Indeed, the following companies contributed some portion of money or services to the project; Adams Excavating, Armstrong Linoleum, Concrete Form Setters, Contact Drywall, Cooper Insulation and Roofing, Don Esch, Inc., Floor Craft, Hendrickson Excavating, Inc., Honeywell, Inc., Mobile Premix Concrete, Inc., Oldach Building Products, Overhead Door Co., Don Porter Plastering, Ready-Mix Corporation, Riviera Products, Inc., Rock Wool Insulating, Rogers Electric, Inc., Starks Furniture, Transit Mix Concrete Company, Vali Hi Painting and Decorating, Waterproofing Service Company, Williams Insulation Company, Wright Glass Company, Douglas M. Jardine, Engineer, and Wright Plumbing and Heating Company.

Despite all of these contributions, the house itself, including land and fees,

"Cost of solar system only $11,122.40"

"There were some contributions"

2,190 square foot home for $95,400

was priced at $95,400.00 It is a 2190 square foot building not including an unfinished family room. It was constructed between March and May of 1974. Colorado building costs, including contractor profit and land, will average to the retail consumer between $22 and $25 per square foot of living area. This would make the home with a conventional forced air heating system **sell** in the $48,180 to $52,560 price range. (The land was $6900, typical for residential lot costs).

Deduced cost of solar heating system: $42,840

The **additional** cost for the solar heating system thus would be from $42,840 to $47,220, **plus donated materials and labor!** For a 780 square foot collector, that works out to between $55.00 and $60.50 per square foot of collector for the entire system installed. That, by the way, is a **typical price range for hydronic type solar heating systems.**

"Mass production" cost: $6,000

An alleged spokesman for the group, falling into the limelight, quoted the $11,122.40 number as if it were the **price** of the system, and then went on to suggest that the same system could be installed on other homes, **with mass production,** for only $6000. Yet everything on the materials list, with the exception of the controls, was already a mass produced item!

85% solar heat proclaimed

Articles have appeared proclaiming the huge success of the Phoenix house system, with the claims that solar energy provided 85% of the total winter heating requirements.

Report to NSF states otherwise

This PR enthusiasm is scarcely met in the January '75 report from Mr. Phillips to the National Science Foundation: "The heating system is at present off-line pending the repair of collector leaks. The problem of obtaining a tight fitting connection between the panel and hose from the header pipe has persisted. We are at present wiring all connections as to the specifications outlined in the 10th November 1974 Monthly Report. **It is projected that the system will go on-line 9th January 1975.** At present the backup system, i.e. heat pump and resistance heaters located in the heat pump, are operational. **The backup systems have been extensively tested and satisfactorily met the heating requirements of the house."** (Heavier type is the author's)

Therefore, if the system did go on-line on January 9, 1975 and provided **100% of all heat required** for the rest of the heating season (dubious), it still could have provided no more than 56% of the winter heating requirements.

Such distortion may be attributable to the participant's laudable desire to make solar heating seem practical and affordable. Unfortunately, such things have a way of getting out — a contractor, for example, **knows** what it costs to build a home — and the result is to give the concept of solar heating a bad name.

Such "successes" then serve as the foundation for further "successes". **U.S. News & World Report,** in a June 2, 1975 article, "Colorado Shows How To Put Sun to Work" quotes Mr. Douglas Jardine, the consulting engineer who designed the Phoenix house system, as follows, 'Next in line', says Mr. Jardine, 'will be a neighborhood solar energy heating system'. His plan is to

No more than 56% possible

Perhaps a desire to make solar seem practical?

54 more "successful" solar houses?

heat 54 low-income homes from a single central bank of solar collectors. **For this ambitious project, Colorado Springs is requesting financial assistance from the Federal Government.** If the money is granted, work will begin next year. (Heavier type is the author's) If the systems are installed at only one-half the cost of the last project, we, the taxpayers will be forking over $1,156,680.00. (And that is solar equipment only – not the houses). I find it particularly ironic to call those "low-income" homes.

The myth of mass production

To return to the revered "mass production" just a moment: Pittsburgh Plate Glass is mass producing collectors today at $11.11 per square foot FOB the factory. With a 720 square foot collector, the cost for this component alone in mass production is $8000 plus freight, plus installation, plus the additional cost to frame in a rooftop at a 55-60° angle. Glass, aluminum, copper, insulation, controls, motors, fans, pumps, tanks are **already** in mass production. How does "mass production" in the future bring the price **down?** One should, instead, shoot straight, and say, "No, they won't come down, friend. They'll probably keep going **up,** just like everything else." The other area unique to most of these rooftop systems is this: they are custom tailored to the house. There is no way to mass produce a custom tailored installation.

Ivory tower unreality

Is the general air of ivory tower unreality present in the field of solar heating beginning to convey itself?

CHAPTER THREE

They walked through the aspen, the boy and the old man, the sunlight behind the leaves making the whispering trees a stained glass window around them. Green and gold and yellow patterns on the rough brown white of the columns holding it all in place.

The fallen logs were no obstacle here, as they had been when they worked their way through the pine forest earlier. The wind had sounded like a waterfall there, roaring mysteriously quite near, yet moving before them like a rainbow.

They rested, finally, when they came out of the aspen into a small meadow, sitting in the afternoon sunshine. Small puffs of steam rose almost unseen from their trouser cuffs.

"Do they still follow us, you think?"

"They're still behind us, boy. But they must work harder in the following than we do in the leading."

"And if they catch us?"

"They will want to steal from us what can't be stolen from any man."

SERVANTS OF THE PEOPLE

Every man **must** harbor in his secret thoughts a dream of personal and total independence – a world without bills, sufficient food, and work with a real, satisfying, **accomplishment** inherent in it.

Personal independence for mankind

Until the practicality of harnessing the sun became apparent, this was impossible for most men. Energy and food in sufficient amounts were out of reach of the ordinary man living in the twentieth century society, without laboring long, often at stultifying and mind-deadening work to obtain the medium necessary to purchase these things.

Harnessing the sun portends such independence

A syndrome of unidirectional thinking led man into unwitting societal enslavement. For thousands of years, wood was the fuel which supplied man's energy needs. Advancing technology added coke, charcoal and coal, and the die was cast. A certain type of thinking was forged. Instead of developing a technology to harvest sunshine, efforts were bent toward finding more efficient **fuels**. First oil, then gasoline, then the ultimate Pandora's box, the atomic fuel.

Syndrome of scientific thinking: find fuels

This "blinders-on" approach created, and over a number of years developed, a new power hierarchy – the energy supplier. It was appropriate in the

New power hierarchy— the energy suppliers

19

minds of the citizenry that one **pay** for the utilization of natural resources justifiably belonging to **everyone**.

You pay for fuels

Certainly it was rational to pay the man or company that expended capital and time to discover oil or uranium for the resources which were on occasion found.

Energy suppliers control peoples and governments

Unfortunately, giant corporations controlling energy all over the world grew larger and larger, finally achieving, today, a control of governments internationally, a control of world leaders and world policy absolutely without historical parallel.

Vertical and horizontal integration expanded control

The fact that these multinational interests integrated horizontally via joint ventures, mergers or other such devices, with entry into **all** energy supply areas — oil, coal, gas, oil shale, geothermal, solar, wind, and nuclear — while simultaneously integrating vertically, by controlling all levels of distribution went largely unremarked until just recently — at least in any tangible way.

Senator Abourezk introduces anti-monopolistic legislation aimed at energy suppliers

Only in January 1975 did a legislator, Senator James Abourezk of South Dakota, introduce legislation which might begin to check this uncontrolled growth of economic power over people and other governments. He introduced a bill January 29, 1975 (S489) entitled, "Interfuel Competition Act of 1975", which would make it illegal for companies to participate simultaneously in more than one of the following: coal, oil shale, petroleum or natural gas, uranium, nuclear reactors, geothermal steam, or solar energy. Remarkably, for such a bombshell of proposed legislation, this controversial proposal received very little media coverage. Yet it is a trust-busting proposal of unparalleled magnitude. Even though it doesn't aim at vertical integration — the vertical integration which forced unbelievably large numbers of independent small businessmen out of business during the past two decades, in particular the independent service station operator, whose plight and demise seemingly went unnoticed and unmourned.

Despite the above bill, which probably stands little chance of affirmative action in the Senate, an even more disgusting power grab is now being nurtured and nourished by the energy cartels and their willing or paid accomplices at all levels of the government.

Even more disgusting power grab being nurtured

In seeming enlightenment, the Atomic Energy Commission was recently legislated right out of existence, to be replaced by a new agency, the Energy Research and Development Administration (ERDA), with the idea that the abuses by the former could be discontinued. The new agency was made up of officials of the former, an act of sophomoric optimism concerning the motives of the involved participants.

AEC replaced by ERDA

I share the suspicions expressed May 18 of this year in the Albuquerque, New Mexico **Journal** in an article entitled, "ERDA's Sun Symbol Inappropriate". It is indeed suspicious when the sum total of ERDA's solar funding is $57.1 million and nuclear funding requests equal $1633.9 million.

"ERDA solar requests— $57.1 million nuclear requests— $1.6 billion"

21

Funding 92% to big business 8% to small business

"Suspicion that solar would not gain rapidly"

G.E. and Westinghouse predict 2% to 4% solar utilization by the year 2000

A.D. Little study arrives at similar conclusion

Dr. Jerry Plunkett of Materials Consultants, Incorporated of Denver, Colorado recently announced that the governmental funding for solar research thus far has been inequitably distributed: 92% to big business, 8% to the small business.

And what was done by the big businesses which received this governmental largess? Senator Gaylord Nelson of Wisconsin said, "The suspicion was almost unavoidable (referring to General Electric and Westinghouse) that these giant firms, because of their large investments in nuclear technology, hoped that solar energy would not gain rapidly." He cited studies by G.E. and Westinghouse, financed by $500,000 grants each by the National Science Foundation, predicting that "within the next 25 years solar energy would be providing only 2 to 4% of total (U.S.) heating and cooling needs, when nuclear energy—a far more complex technology—had jumped from zero to 6% as a source of electrical power in less than 20 years." (May 20, 1975, **Christian Science Monitor.** "Builders Say It's Ready Now.")

The A. D. Little Co., in a study financed by a large number of corporate giants, not very surprisingly, arrived at a similar conclusion! Unfortunately, non-engineer public officials are duped by such sources. They treat the conclusion as a pronouncement from the oracle at Delphi!

Another quote from the same **Monitor** article, "Raymond D. Watts, general counsel of the Senate Small Business Committee states, 'Nuclear technology is big business technology, whereas the small business is uniquely equipped to develop solar heating and cooling hardware. The power establishment is dragging its feet, because if we went too far, too fast (on the development of solar energy) the disruption of established technology would be too devastating!' "

"Small business might disrupt nuclear technology"

I found these latter remarks by Mr. Watts of particular interest. I met with him and some of his assistants at the Sheraton Park Hotel during the ill-fated SEIA-FEA-ERDA Expo 75. Despite the above comments concerning the power establishment foot-dragging, he asked me about reports that "your unit is collecting more energy than falls on it". Again, the source of these reports was undisclosed. With Mr. Watts was an engineer, Ron Larsen, who stated, at least, that he believed that the backyard furnace could achieve the performance claimed for it. I suspect a naivete in Mr. Watts. I think that he can visualize "foot-dragging", but not make the next step in logic, that maybe some downright interference (via discrediting techniques) in the successful introduction of solar energy utilization could **also** be occurring.

Meeting with Mr. Watts at Expo 75 "Backyard unit too small?"

Other suspicions concerning the new ERDA have been expressed. In a June 1, 1975 article in the **Philadelphia Enquirer** entitled, "Sun Power Clouded by Fear", Representative Ottinger states, "ERDA is just the AEC glazed over with a new name and has few people on the side of solar energy." Adds an aide to Congressman Charles Vanik of Ohio, "The government is not about to throw

"ERDA just the AEC glazed over with new name"

away investments in atomic energy, and as a result is unconsciously neglecting solar power."

Something worse

I disagree with that last statement. ERDA is certainly not "unconsciously neglecting" solar power. Something **far worse** is brewing...

CHAPTER FOUR

The leader of the group stood impatiently waiting while the scout hunckered near the ground.

"How far are we behind them?"

"About six hours, maybe more."

The thin one shifted uneasily and moved to face the leader.

"Mebbee it ain't worth it. Mebbee they're right."

"Our way is right. That's what you get paid for. The group has decided, anyhow. It's not my place to change all that. Nor yours."

The scout had moved on up the slope, walking hunched over like the gnarled trees that spotted the mountain top in defiance of the raging winds.

The men followed.

ENERGY BLACKMAIL TO FORCE IMPLEMENTATION OF NUCLEAR POWER

The Solar Energy Industries Association — FEA-ERDA Expo 75 held in Washington D.C. on May 27, 28, 29, 1975 was an orchestrated musical comedy. One of the scenes:

Expo 75 a musical comedy

Congressperson Mike McCormack of Washington State is known as an outspoken advocate of nuclear energy. To quote the press release handed out at EXPO 75, "He is the leading Congressional advocate for the Nuclear Fusion Program, and has obtained substantial increases in funding for this program for Fiscal 1974, 1975 and 1976. He is the sponsor of legislation allowing greater participation by individual states in the siting of nuclear power plants and a plan to create a **nuclear power park** in each region of the country." (Heavier type mine. I find the use of the words 'nuclear', 'power', and 'park' in conjunction with each other a massive statement) The release continues, "In addition to his other duties, McCormack was recently appointed chairman of the Environment and Safety Committee of the Joint Committee on Atomic Energy, and of a special subcommittee to review the National Breeder Reactor Program."

Congressperson Mike McCormack foremost advocate of widespread nuclear implementation

This is the same McCormack who "was **critical** of Ralph Nader, who has been actively lobbying for a moratorium on

McCormack critical of Nader

nuclear energy and who has been telling members of Congress that solar and geothermal energy can solve the nation's problems," in "Energy and the Environment" **Rural Electrification,** May 1975.

Super Nuke named Super Solar

In what should receive an annual award from the **Harvard Lampoon** as the most devious political move of 1975, that Congressperson Mike McCormack was given the "Man of the Year Award" by the Solar Energy Industries Association! The man known as "Super Nuke" in environmental circles had received an award as "Super Solar", and few of the SEIA members present saw any contradiction in that! In fact, to add debasement to degradation, these members followed the lead of the FEA and ERDA officials present in giving McCormack a standing ovation **after** he gave an acceptance speech predicting 1% solar utilization by 1990, right in line with the nuclear club's predictions.

Award for National Solar Demonstration Act

The ostensible reason for granting McCormack the Solar Man of the Year Award was the fact that he was the prime sponsor of the National Solar Heating and Cooling Demonstration Act of 1974.

Solar legislation designed to hinder solar

Let's examine this "landmark legislation". If ever a bill was passed to impede the progress of solar energy and to hinder the implementation of solar heating, this bill was an epitome of such.

$50 million to build 4,000 solar homes

Here's how it works. The government will allocate $50 million to build 4000 demonstration homes heated and/or cooled by solar energy across the U.S.

to "encourage utilization of solar energy". The National Bureau of Standards (NBS) was directed to formulate criteria for evaluation of such systems. Innocent and praiseworthy, right? Wrong.

Somehow, National Aeronautics and Space Administration (NASA) got involved, as did Housing and Urban Development (HUD) in the formulation of parts of the "criteria" and NBS dutifully whipped up the most uncoordinated, irrelevant piece of garbage that ever masqueraded as a work of scientists. Not even the Flat Earth Society has ever been so creatively absurd.

Anyone working in the field of solar energy applications knows that solar collector operation is a function of the interface with the storage component, the heat transfer methodology, the controls utilized and the sizing provisions made to interface the device with a home. Yet the NBS draft interim criteria proposed were unbelievably complex with no reason therefore; were vague and ambiguous, indicated a clear misunderstanding of solar applications in general, and included the collection of enough irrelevant data to require the services of thousands of scientists for years to perform the tests suggested. One wit decided that NBS had been instructed by NASA to find jobs for **all** of the aerospace engineers who were going to be unemployed with the demise of aggressive space projects. None of the NBS tests would have resulted in useful information for evaluating total systems, although, in fairness, there was someone working on the project who slipped something useful into the tests when no one was

NBS to formulate "criteria"

Complex component tests that yield little or no relevant data

No useful information for evaluating systems

29

looking – and that pertained to life expectancy testing.

The worst and most virulent intent, however, was made clear when the criteria were unveiled to public scrutiny in a special meeting in Washington, D.C. on November 21, 1974. A full-blown PR type presentation was given by dignitaries of NBS complete with beautiful, colored slides. The slide presentation contained in several places, the word "certification". The dignitaries also referred to "certification of systems."

Malevolent intent— "certification" by NBS of privately manufactured systems

Clearly, governmental regulation of the new solar energy industry was intended and was being planned. Later legislation concerning solar incentives bore out my worst fears. The words "for certified systems" are attached as conditions for qualification for those incentives in much of that proposed legislation.

Governmental regulation of private sphere intended

International Solarthermics Corporation launched a major offensive in November 1974 objecting loudly to this unwarranted intrusion of government into free enterprise. What other furnaces does the National Bureau of Standards certify? Lennox? Climatrol? Fedders? Trane? General Electric? Nope. Yet surprisingly, none of the giants involved in solar energy research offered **any** objections. Why not?

ISC objected to intrusion into free enterprise

Such a "testing program" **had** to significantly **impede** solar energy introduction.

Testing program would impede solar energy implementation

What has happened in the interim? Despite the fact that hundreds of small businessmen from coast to coast are **selling** at retail, solar heating systems for $3000 to $6000 each installed, a NASA analysis showed that only 650 (instead of the originally projected 4000 systems) could be installed for $50 million! More. An additional sum of $37 million would be needed for test instrumentation! That works out to a cost of $76,923.08 per system installed, plus $56,923.08 per installation for test equipment. Who were the potential suppliers consulted for the NASA study? Certainly not those small businessmen already selling units in the marketplace. For they could have supplied over 10,000 systems for the $50 million, not 650 as predicted!

The insidious part of the Demonstration Act lies elsewhere, however. (As if $125,000 per heating system and test equipment to demonstrate its "practicality" were not enough!)

Picture a small businessman selling tested systems with a guarantee of performance in Smalltown, USA. He talks to a customer who is contemplating the purchase of a solar furnace. The next morning the customer reads in the newspaper that 650 systems are going to be installed and **tested for five years to evaluate their performance.** The customer is no dummy. He reads, correctly, that the government is going to install 650 systems for five years **to see if they work.** What is the customer going to think of that honest small businessman's statement, "I can, after I perform a heat loss calculation on your home, tell you how much of your heating requirements the furnace will provide, **before you buy the furnace?**"

NASA reports not 4,000, but only 650 solar furnaces for $50 million

$76,923 for each solar furnace to demonstrate practicality?

Insidious part lies elsewhere

"Put them on houses for 5 years to see if they work"

The Act discredits engineers by implying they can't calculate performance

The customer is a layman. He doesn't understand that even a first year engineering student can measure Btu output of a solar furnace, and, using years of weather data, can predict very closely the performance of a furnace on a given structure. Thus, the real world impact of the Demonstration Act is to discredit the honest engineer by implying that the only way you can tell if a system works is to put it on your house and wait five years.

Act disastrous to introduction of solar energy

Congressperson Mike McCormack's National Heating and Cooling Demonstration Act is disastrous to the introduction of solar energy utilization, impedes its spread, and is harmful to the honest men working in the field by discrediting what they say to their customers.

"Solar heating is untried and mysterious"

The implication rings loud and clear in the mind of the average American, "Solar heating is new, untried, mysterious." Was the Act an honest attempt to forward the cause of solar energy or a callous political move designed to accomplish the opposite?

Governmental solar projects outlandish, gawkish, cumbersome and impractical

The many grants being disseminated are also ostensibly for the purpose of forwarding the cause of solar energy utilization. Yet the projects would have given Rube Goldberg cause for envy. Even his most outlandish creations had a sense of order within them. The solar projects, on the other hand, have been outlandish, gawkish, cumbersome and impractical. There is widespread agreement by manufacturers of solar heating systems in the United States that from a cost-effective standpoint the only practical system for

32

single family residences is the forced-air type solar heating system. The hydronic (water transfer) type system costs many, many times more for an equivalent useful heat delivery. Yet at least 99% of all governmental funding has gone to the hydronic type systems. (Indeed, the 650 homes proposed under the National Demonstration Act would cost so much only if they **were** hydronic).

How better to alienate the marketplace and impede introduction of solar than by building gargantuan gruesomes which would crouch over the entire house and reach deep within its bowels somewhat like Hollywood's space spiders? How better, graphically, to demonstrate the impracticality of solar energy?

Alienate the marketplace with ugly systems

Yet another type of subtle discrediting of solar energy goes on. At taxpayer expense and smacking slightly of P.T. Barnum, Honeywell has a traveling circus on wheels, sporting a large and ungainly solar collector, which travels from city to city "collecting data on solar energy". The extent and duration of such data collection makes its usefulness dubious in the extreme. One can't help but note, however, that unusual amounts of press and television coverage accompany the arrival of the van in most cities, and that huge amounts of useful PR collection **are** occurring. One can be certain, of course, that that result was the furthest thing from the minds of Honeywell management in their altruistic proposal to further the cause of solar utilization. Strangely, Honeywell is still missing from the ranks of the businessmen who

Solar traveling circus— promoting company at taxpayer expense

have invested their **own** money in marketing solar furnaces throughout the United States. One should be permitted at least to suspect that when the publicity is no longer forthcoming, and when the pioneering businessmen have established a stable marketplace, why then, and only then, will Honeywell **risk** an entry into the marketplace!

Implicit derision of concept

As you read solar energy stories in the future, examine this phenomenon. Is the project one which only J.P. Getty could afford? Is the house aesthetically a nightmare? Is there a subtle sense of the absurd? If so, be sure that it is a project designed to impede, not forward the cause of solar. Notice how little the sense of "practical", "useful", "inexpensive", "affordable" impinge in those articles. Implicit derision should become apparent to you.

Discredit the practical systems

What happens to a **really practical** system? Impede it any way possible. Probably the easiest way is a whispering campaign. One emanating from a nebulous "authority" who disdains to identify himself. (For good reason — libel and slander suits could be really embarrassing). I have seen this technique in operation personally, and can testify that it is the most difficult of all to fight.

More on the "big lie" whispering technique later.

CHAPTER FIVE

The boy was angry, frustrated. "Why do we do this, old man?" They had waited, now, for over three hours, watching the backtrail, hidden from long view by a massive spruce, but seeing the long slope lying in the sunshine below them.

"We must stay within reach, boy."

"But why?"

"They'll be getting discouraged about now. We've got to keep them together, so the others can see them together."

The old man pulled a piece of paper from his pocket, and scribbled a few words upon it. Weighted it in the middle of the trail with a rock.

"They just crossed the ridge. We can go on, now."

The boy joined him stiffly, one leg had gone to sleep. A chipmunk watched them nervously out of sight.

2058002

ENGINEERING HUMBUGGERY

Although it is conceivable, and perhaps even probable, that the issue to be discussed in this chapter is part and parcel of the general attempt by the nuclear club to portray solar energy utilization as totally impractical, two other alternatives exist...

Poor engineering may be part of discrediting technique

First, that the engineers involved are simply missing typical engineering design considerations of very elementary nature, i.e. that they are incompetent, or,

Second, that the engineers involved are stupid.

But engineers may only be incompetent or stupid

I have looked at any number of solar collectors, of the flat-plate type, using water or air as the transfer medium. A very interesting similarity pervades most of these. There is a total lack of attention to **conserving** the precious energy collected at such an expenditure of materials, time and money. The temperature inside a collector, when it is operating, will be from 100° to 250°. 30°to 180° hotter than the air inside your home. Yet 9 out of 10 of these solar collectors will have only 1 to 2½ inches of fiberglass insulation!

Little or no conservation of energy in new designs

If you bought a new home with only 1 inch of insulation in the attic, you would undoubtedly sue the builder. Yet instead of more insulation in solar collectors than you have in your attic,

Only 1 inch of insulation in your attic would be cause to sue the builder

because the temperatures are much hotter, one finds **less!** I call that spurious, or at the very least lackadaisically incompetent engineering. If it is a glass company selling the poorly insulated collector, one may easily infer a motive: to sell more glass. The harsh judgment of that type should not be easily made, however, for perhaps the engineers involved were just stupid.

The problems with conclusions of stupidity, however, or even more run-of-the-mill, mediocre incompetence is that even a first-year engineering student wouldn't make such an elementary design error let alone a professional design engineer. A further problem exists with either of the harsh conclusions, because tooling up to produce collectors costs money. It is hard to imagine that a hard-headed management group in charge of deciding to enter this new field would knowingly take the risk of involving their company in a marketplace in the pioneering stages with an obviously inferior product, since an honest competitor would soon make mincemeat of the company with a good product. For the competitor would not waste time pointing out such obvious deficiencies.

So although it would be neat and simple to level a finger of judgment and condemnation, opting for the conclusions made above, more than likely neither conclusion is valid.

So what is the answer? Why is such shoddy design work being done?

Probably the answer lies in the history of the development of solar heating applications in this country.

38

Very little practical, innovative work has been done until recently with solar energy in the U.S. Dr. George O. Löf and Harry Thomason are the real pioneers of solar heating in this country. Both worked by themselves, at their own expense, back when the concept of solar heating had no popularity or real need in the minds of the populace during the late forties and fifties. Like most innovations, their work was developed in the true tradition of backyard tinkerers and hobbyists, with a lot of make-do, because the right thing wasn't at hand, or wasn't affordable. Undoubtedly both were regarded as crackpots and cranks by their neighbors. Each developed a system of his own quite different from the other in overall approach. Most of the systems you see today can be traced directly to one or the other of these early systems. They made "mistakes" of course, in engineering. But with 20-20 hindsight, to focus on those "mistakes" would be to miss the point – inadequate insulation on those collectors was unimportant – the goal they had was to prove solar energy utilization **possible**. And they did that. Some of the systems they built back then are still functioning today.

Unfortunately, the work that followed was largely imitative. The "mistakes" – the details of careful engineering which had been overlooked in the quest for the larger goal – were repeated without thought until they became "the way to do it".

Then, a couple of years ago, the world woke up to what Löf and Thomason had seen so long ago: our precious fossil fuel reserves were being depleted

Answer lies in tradition and history of solar applications

The pioneers' goal: to prove solar heating possible

Work that followed was imitative— even of "mistakes"

39

alarmingly fast — what were we going to do?

Energy crisis precipitates big time science approach

Suddenly, big time science became involved in solar heating. Big time science and an academia conditioned by the recent trauma of aerospace approaches to the problem of putting a man on the moon before the Russians did. A kind of approach to discovery in which the high technology involved was so complex that **extreme** specialization was required, giving birth to team science, in which the problem was piecemealed to death. An approach where the orientation was toward obtaining esoteric solutions — Buck Rogers ideas. Most importantly, an approach where, due to the crash deadlines of a race, **cost was no object.** In that high pressure environment, the scientists lost the concept of cost-effectiveness.

The specialists never saw the system as a whole

So a specialist transferred from working on radiation problems in outer space to working on a solar collector just wasn't concerned about the amount of insulation (prosaic, basic stuff) on that collector. He was concerned about the reduction of long-wave radiation from a collector plate surface, by using specially formulated (and very expensive) selective coatings. Working in the laboratory with his specialized problem, he never got an opportunity to learn that collector plates get **dusty** in use, thus obviating the importance of what he was doing.

High technology approach without coordination

In short, a high technology, specialized, diversified, components-and-parts-of-components approach was implemented without even the kind of overall coordination that existed in the aerospace effort, which might have been a saving grace.

If even one of these scientists had studied in breadth the problem of solar energy applications, he would have seen that collecting solar energy, due to its lack of concentration in any one place, is a **low** technology science — a science of capturing solar energy with a solar collector, sure, but more importantly, a science of **conserving and storing that captured energy**. Conservation of energy is not nearly as romantic as collecting it, admittedly, and probably that is why it was never given adequate attention. Cost-effectiveness is also not a romantic concept — one's picture of the visionary scientist pioneering the future does not include the prosaic and mundane attention to crass dollars and cents trade-outs.

Solar applications are low technology

The solar research effort grew like topsy without adequate and evaluating guidance. Numbers of competent specialists took portions of the problem without any relation to the overall goal — to design a practical, working inexpensive **system**. Their efforts were often duplicative, or irrelevant.

Many efforts duplicative or irrelevant

As a result, one sees excesses in one area of design, inadequacies in other areas of the same system. One lavishly funded governmental project, for example, used **triple-paned** windows in the house, but had in that same house, heat-wasteful cathedral ceilings and only 4″ of attic insulation in parts of the house and only 6″ of attic insulation in other parts. And, of course, the solar collector and storage tanks were woe-

Excesses and inadequacies in the same project

fully inadequately insulated.

Basic homework was ignored

The solar experts quickly gained the ego of expertise. Because studying heat loss from a home was a demeaning, everyday kind of thing that "even a furnace contractor" could do, it was ignoble for an expert to perform such rudimentary calculations. One solar "expert" has toured the country for years, stating from the podium that "it takes a million Btu's per day to heat a home." (Where? Atlanta or Green Bay? The fact of the matter is that even a typically insulated 3 bedroom tract home in icy Minneapolis only requires an average of 306,000 Btu's per day during the heating season.) But this "fact" was intoned from the pulpit often enough that many researchers accepted it without question. I have often been pompously informed that the backyard solar furnace just can't do the job of heating a house, because it doesn't produce a million Btu's per day, even on a sunny day! Right.

Growth of a new mythology of expertise

This same solar expert produced "truisms" prodigiously. Another ascribable to him was this "rule of thumb": "It takes a solar collector half the square footage of the home to heat it." This one was **generally accepted** despite its apparent ridiculousness. The joke of 1974 in the solar community was the fact that the National Bureau of Standards solemnly included this bit of gospel in its criteria!

"Rules of thumb" accepted without question

The promulgation of such bits of wisdom from the solar "experts" causes furnace contractors, in particular, to form a very poor opinion as to the competency of the research being done in the field. They know that the heat loss from two 1000 square foot homes sitting next to each other may be radically different. One, for example, may have more door and window area. One may be square, the other L-shaped. One may have ten foot ceilings, the other eight. One may have a slab floor, the other a crawl space. One may have decent insulation, the other may not. And certainly a home in San Diego doesn't need the same furnace that one in Helena, Montana does. Only by a physical survey and measurement of the house can a heat loss calculation be performed. And that calculation must take into consideration just how cold it gets outside in the winter! It's bad enough to have such accidental discrediting of the field of solar utilization by well-intentioned "experts" but one must shudder at the thought of the National Bureau of Standards self-importantly undertaking to **regulate** this new industry with such "rules of thumb!"

These bits of wisdom discredited solar expertise

Such dissemination of misinformation is not peculiar to solar energy. If one looks back a bit, he sees that such controversy occurs in the innovative stages of any new technology, particularly as it comes to market.

Such misinformation typical in innovative stages of new industry

After such a lengthy diversion from the starting analysis of this chapter, we must then return to yet another possibility to account for the rotten design occurring in the field of solar collectors. It is possible that no ulterior motives exist, that no downright stupid engineering is going on, but rather, that **no** engineering is going on—that the design is simply the result of unthinking imitation.

Perhaps not incompetent engineering, but no engineering

43

CHAPTER SIX

The scout found the note beneath the stone. Scrawled and irregular, written with a pencil on paper stretched over a knee.

"Why are YOU helping them?"

All it said.

He looked at it, perplexed. What kind of message? He waited for the rest of the group, contemplatively. Gave the note to the leader who read it quickly, crumpled it and threw it on the ground.

He looked at the scout, expectantly.

"How far are we behind them?"

"Hour or so at most. They waited here quite a while. I don't figger it. Cut their lead by an awful lot. Neither of them is hurt. No sign."

"Who knows? Crazies anyhow."

The scout looked away, and reluctantly began to move to the west again.

The sun was getting red, low in the sky. It would be dark soon.

DAVID AND GOLIATH

Thus far this treatise has presented the giant energy suppliers as evil, maleficent ogres clothed in four hundred dollar suits, clustered about polished mahogany board tables, cunningly plotting the eventual enslavement of all mankind to their profit making schemes.

Giant energy suppliers presented as evil ogres

Indeed, the multinational corporation has undergone some avid scrutiny in the recent past and certainly the **results** of the multinational operations have been to increase poverty levels — "The rich get richer" — despite vigorous espousals by management of the same corporations of goals quite the contrary.

Results of giant corporations' control have been bad

Although one may be certain that all is not magnanimity and enlightened humanism in the major corporate boardrooms, I would like to present an alternative picture — one of runaway bureaucracy, sheer bigness, operating to the detriment of the big corporation.

Result of evil management or runaway bureaucracy?

Perhaps I am being rosy-eyed, but with your indulgence I will yield to a possible different interpretation.

First an examination of the vantage point from where I speak.

First, a bit of history

I am a native Coloradoan, rare beast, fortunate enough to have designed with some other men, built and filed for

A-frame backyard solar furnace practical solution to auxiliary home heating

patents on a very practical, A-frame shaped, backyard solar furnace, which was compatible to existing homes. A company was formed to ready the product for market, perform extensive verification testing series to confirm the testing results of the early inventors and be sure that no errors had accidentally occurred. Two very high calibre engineers were added to the staff of that company to test and retest the furnace, refine production methods, reduce costs and make the unit market ready...

Market too large for any one corporation

After several false starts, it was decided that the market was incomprehensibly large, and that only by a massive involvement of many corporations nationally, **big and small,** could widespread utilization of solar energy be accomplished. Therefore, a program of simple licensing under patents pending would be used — an inventor's royalty agreement, in other words — and that the research and development facility would **not,** under any circumstances, manufacture or sell solar equipment. A small corporation in the mountains of Colorado had neither the expertise nor the capital to embark on such an adventure.

Involve many independent corporations operating without restriction

The licensed manufacturers would **each** have the right to make, use and sell the furnace all over the United States **in direct competition with all other licensees.** No more than 100, and perhaps less, would be licensed, but each would have the right to sub-contract manufacturing and distribution. The only exclusivity that the licensee would have would be gained through trademarks, changes in the cosmetics

48

of the unit, improvements in design and quality control, and the business ability to bring production costs and hence price down to give a stronger competitive edge in the marketplace.

In short, an anti-monopolistic, traditional free enterprise approach to introducing a totally new industry to America.

Giant, mid-size and smaller manufacturers were approached with this proposal. Before long, they were approaching us, in the historic "better mouse-trap" syndrome.

It was here that a tremendously interesting and vital fact emerged. The small and mid-sized manufacturers were only marginally concerned about competition from the "big boys". But the major-sized corporations did not like the invention licensing approach at all!

Why? Representatives from two of the bigger corporations finally cleared up the mystery in very plain talk. The smaller companies were planning to **retail** the solar furnace for three to three and half times the cost of materials. The big companies, according to those representatives, however, need at least 6 times and preferably 8 to 10 times the material costs mark-ups in order to show a profit.

With appropriate marketing and advertising techniques, the giant companies could still have operated successfully in the open market, had the product been a different one.

The second problem, the real clinker from their standpoint, existed. The cost

Anti-monopolistic free enterprise

Small businessmen liked this approach. Big business did not.

Small business needs much less mark-up

of the solar furnace itself. Why? Assuming the cost of the materials in huge quantities at factory direct prices to be $1500, the small manufacturer could sell at a profit through a standard distributor-dealer network at a retail, installed price of $4500. The major manufacturer, on the other hand, **had** to sell at retail for between $9000 and $15000. Granted, the typical monopoly approach to competition of lowering prices and forcing the smaller competitor out of business would still seem to be appropriate here. Not this time. A loss of from $5000 to $11000 **per unit** was unthinkable, even if all competition could be forced out of the market in a very short time. For every 100,000 units sold, that represented a **loss** of from $500 million to $1.1 billion! Even a giant corporation could not sustain such huge losses.

This is an exciting and new precedent —one which reverses all of our thinking and mythology concerning the tremendous advantages of assembly line production, and the cost advantages of assembly line production. It just ain't so, in every case, anymore.

The macroeconomics and macromanagement of giant corporations have made them contraproductive.

The significance—the major importance—of this discovery was lost on me at first. But finally it dawned. The age of the assembly line is over—at least for existing major corporations producing any but the very inexpensive, small product.

The shocker that finally woke me up will wake you up, too. As part of my research into producing a solar-powered automobile, I had occasion to ask a small, local custom auto maker to price out for me (using fiberglass bodies instead of steel) the components necessary to the making of a major sized American automobile, except for the engine, but including everything else, from bucket seats and padded dash, from seat belts to chrome wheel covers. He was to get as many components as possible package-assembled, so that assembly problems could be reduced, from the same suppliers who supplied the Big Three. The small quantity purchase OEM (Original Equipment Manufacturer) price on this total package was astounding. A mere $900.

Now, I reasoned, if I could get a complete $6000 to $7000 automobile for $900, what are the Big Three paying for the same thing? Perhaps $600? I don't know. And one would have to add in the cost (OEM) for the engine. But I do know this: a competent, small assembly plant right in your home town would produce the whole automobile, including labor and reasonable profit, market it through a local, small independent dealer with **his** profit for about $3000 to $3500. So that a small, custom produced automobile would cost about half what an assembly line produced automobile would!

I had bought the American myth – that mass production technology made everything cheaper. There is truth in some parts of that myth, because the assembly line **does** reduce per unit labor costs. But giant corporations have grown out of control into huge, management and stockholder top-

Automobile for $900!

What is Big Three mark-up on auto?

American myth–mass production makes everything cheaper

heavy, cumbersome bureaucracies which require extraordinary portions of the income to support.

Assembly line better for high quantity, very small items

The old assembly line dictum must be modified – it applies to low and moderately priced items with sufficiently broad market appeal to justify high quantity central production. And the items must be small enough that freight doesn't eat any savings achieved through central production. Of course, the modular home builders have been learning this to their dismay for the past ten years – big ain't necessarily good!

Quality control a function of worker pride

The other argument for mass production and big has been that quality control is much better. Not necessarily true. The quality of many products, particularly those involving the assembly of lots of components, is in direct proportion to the pride of accomplishment of the workers assembling it. A recognition of that fact has been dawning slowly in the auto industry, in particular. Experimental work has been done with "team assembly". Notice, however, that the **consequence** of the latter approach has been overlooked, should it be implemented in a more widespread way. If a small team assembles a total unit, what is the necessity for crowding **all** of the teams under one roof at one location? Why not spot small, regional plants closer to the market itself, saving the expensive freight of delivery?

Contra productivity due to growth of corporate bureaucracies

How has this contraproductivity arisen? Unfortunately the exigencies of giant corporations require the continual introjection of subsequent layers

of middle management, until management overhead per unit produced may
well exceed the cost of materials and
labor to produce that unit.

Why are the continual additions to
management required? In one word —
communications. Ask any manager of
any business what his biggest problem
is, and if he is aware of his organization, he will point to communication
as the culprit.

Even small companies, with less than
25 employees, must have rigid structure and continual meetings to
overcome the problem of poor communication. Why? The goals of the
company itself seldom correspond with
the goals of the aggregate members.

Even if the direction and goals of the
company are agreed upon, there are
disputes or various interpretations concerning the methodology necessary to
achieve those goals. Thus "communication" is really a label covering a
complex of minute movements and
adjustments by the participants, always, hopefully, aimed at the achieve-
ment of the company goals.

Unfortunately, personal problems and
motivations may affect or interfere with
that performance and subsequent interpretations. In reacting to this problem, as the company becomes a giant
employing tens of thousands of people, the management is forced to continually add communications (read
more managers) to their staff. As
authority and responsibility are di-
luted and delegated further and further, the bureaucracy grows. At some
point decision making becomes dan-

Communications require added layers of management

Secure agreement of employees goals with corporate goals

Authority and responsibility diluted— new decision makers continually required

gerous to the decision maker. and the continual introjection of new decision makers is required to keep any impetus in the company. The added decision makers further dilute responsibility and the cycle continues. Flexibility is totally and irreparably lost.

Conservative management prevents progress

The founders of the major auto-making corporations must be turning in their graves at the kind of "management" existing in their companies today, not to mention the wages they receive. In response to a governmental demand for cleaner engines, their response surely would not have been to garbage up the existing engine with additions of layers of cumbersome and inefficient equipment which further reduced the efficiency of the engine. Yet the modern "management" has done just that; perpetual increases in the inefficiency of the product. Where the obvious solution was to invent a better, cleaner engine.

Anti-pollution equipment on automobiles a joke

The ordinary man is not fooled by this corporate obfuscation, despite the "tremendous impact of advertising today". Any man in the street can tell you that it is the anti-pollution equipment hiding his engine from view which is responsible for degraded performance and mileage. One must view the squid-like mass of contortions which wriggle leeringly when one opens the hood as nothing short of an engineering obscenity. (Or a practical joke)

Giant corporate bureaucracies rival government bureaucracies in unwieldiness and inefficiency

The giant corporations' bureaucracies rival the government's own in size and unwieldiness. And it is this that is the problem. The bureaucracy becomes amoral or outright immoral. In talking with the individuals caught up in such corporations, one hears the same kind of language universally – what can one man do, who can buck the system – in short the frustration of seeing the need, but being unable to accomplish it. Or – I **need** my job, I'm not putting my neck on the chopping block, I only have five years to retirement – the language of alienation, inadequacy and impotence.

When some crusader approaches an abuse head-on, marshalling support for his crusade, over-compensatory excesses result, due to the panic of the individuals within the corporation scrambling to evade responsibility on the one side, and others subtly directing without becoming responsible in that process, prescribing reactionary solutions. A tremendous see-sawing results, requiring more legal staff as the problem deepens with the complications of the reactions to the crusader, and requiring more on-staff "defensive" reactionary management, and the unwieldly beast continues to grow.

Attacks on corporations require additions to management bureaucracy

Therefore, the point of this analysis is not to paint the giant corporations as evil and inhumane because fiendish and malfeasant members of the board control the monolith.

It is precisely the opposite. Any man sitting in the presidency of a giant corporation (or country for that matter) who thinks he is in control is pathetically deluding himself. He is riding an out of control dinosaur which by its very nature will become more and more uncontrollable in the future. Once the behemoth is adding to its bureaucracy continually like a cancer flourishing in its own suppuration, it has gone far

No man controls the bureaucracy, —it is a mindless beast

past the capabilities of any man to relate to and understand the goals and direction of the company he "runs".

The complexities outstrip man's capacity to comprehend

The recent national debates by the pundits concerning the state of the national economy — was it recession, depression or inflation — indicates clearly the complexity of interrelationships which can develop beyond all capacity to comprehend.

Self-destruction is inevitable in the multinational corporation

Self-destruction is inevitable in the giant corporations, despite the fact that it is terribly hard to acknowledge such a pessimistic (or optimistic!) possibility in today's environment. The flexibility is lost. The brute power available to crush an opponent still exists and is used with unconscionable regularity. But force requires direction if it is to be useful and remain strong. Many times that force is accidently utilized and innocent victims are steam-rollered needlessly. The power will remain for some time to come, but power without intelligence is foredoomed — probably power **with** intelligence is, too, for that matter!

Unrelated decisions are often contradictory

The amoebic aimless movements of the giant corporations are not without certain moments of horrified humor. Their self-destruction is often hastened unduly by the making of separate but unknowingly related decisions.

Hire low IQ personnel and only promote from within

A giant corporation some years ago formulated two policies. One addressed itself to the problem of employee attrition from the daily boring jobs of assembly of the product. Personnel groups and committees were called together, and not unsurprisingly, psychological screening of all prospective

56

employees, including IQ assessment, was adopted.

Only placid, well-balanced, low-IQ personnel should be hired, because they would stay on the job forever, reducing the iterative training expenditures. Nearly simultaneously another group of committees was studying the problems of management morale, and the satisfying decision was adopted that company-wide — all promotions should henceforth be from within. One shudders to imagine the resulting capacity of not just a corpulent bureaucracy, mindless as an organism, but also mindless from within! Now there's a reverse Darwinian constipation of unrivaled breadth.

Reverse Darwinian constipation

Corporations must be kept manageable, if they are to compete, survive, lose occasionally and survive the loss, growing not in size but in capability, because sheer force alone can't insure survival. Witness the dinosaur, the Roman Empire, the German war machine, the Nixon Administration.

Corporations must be kept small enough to be manageable

The small businessman is more efficient, more competitive in the long run, probably because he knows where he is going, and he cares. Recently, in a televised address, President Ford acknowledged the contribution of the small businessman and gave a startling statistic — 62% of our gross national product is attributable to the small businessman. The small businessman contributes more to the country than the giant corporations who control it.

The small businessman cares . . .

And his contribution is very significant

CHAPTER SEVEN

They ate in the darkening. The boy with the haste of youth, the old man more slowly. The pack was getting lighter.

"Have to get more food soon, old man."

"Soon, boy" the old man agreed.

No fire tonight. The followers were too close behind.

"They won't move at night, you're sure?"

"This is a daytime game, boy."

A nocturnal bird, probably a mountain owl, flew by, sensed as much as heard. It was going to be cold tonight. No clouds, moon in an hour or two.

The boy thought of that. "It truly is the sun that lights the moon?"

The old man nodded, reassured by the thought, too. Still there tonight, unseen; still there tomorrow.

THE CONTRADICTIONS —WHY?

Suppose the following scenario: one is loaded down with investments in nuclear technology and high price uni-use equipment, and the market is not moving. How to implement it? How to get the market moving again to regain the investment?

Suppose further that an opportune fuel shortage—a real shortage—is looming in the near future. Obviously if the oncoming shortage is recognized and reacted to by the government and/or the public-at-large in time, the predicted shortage will never develop into a crisis.

Yet in the estimation of the experts on the board of directors, only an emergency footing can stimulate the nuclear market again, only then can the public resistance to atomic plants be overridden by the outcry for immediate power supply solutions.

The problem with this scenario is how to keep the public and the government and the press from cottoning to the imminent emergency too soon and taking steps to counteract the emergency **without** nuclear energy.

Simple obfuscation won't work. You can't hide a problem like a shortage, because it doesn't appear overnight. Rather, regionalized dislocations in supply occur sporadically, as the logis-

The nuclear club dilemma —how to implement faster wider utilization

Fuel crisis looming

National emergency required to override safety objections

Prevent other solutions before emergency status achieved

Prevent action on impending fuel crises

tics of supply become more difficult.

A much more complex strategy would have to be used. Something more devious. One, for example, would never be suspected of discrediting oneself. How difficult would it be for an energy supplier (vertically integrated) to conceal market manipulations? Not difficult at all, from a business standpoint. On the other hand, if one deliberately allowed the curtain to be raised on clumsy manipulations, would not an aura of distrust be created? Would not the general assumption be that the supply was adequate, and the manipulations purely aimed at driving prices up?

Discredit and obfuscate, create aura of suspicion

In the natural public feeling of security and national pride that "we've always been independent, and we can take care of ourselves," is the psychological seed that such a ploy could rely upon. Talk to the average man in the street, and he'll assure you solemnly that we've got enough fuel—it's just an oil company-Arab conspiracy to make profits.

"The fuel crisis is an oil company conspiracy" attitude will hinder positive solutions

Intimate that the fuel reserves have been understated

To nail that move in the scenario down, have some of the people within, imply that the estimates of fuel reserves have been **understated**. Who then would take seriously the National Academy of Sciences when they announce that the Federal Power Commission fuel reserve **estimates** are twice the amount of proven reserves that actually exist?

Develop more alienation

The effect is to discredit all of the prognasticators. Who is the ordinary man to believe? Probably none of them is the natural, frustrated response. And a little more alienation develops.

Notice how we talk about the fuel shortage. It is always in a context of disbelief. We talk about it in the same stilted language we use to talk about **death**. Third person. Denial of personal involvement. Denial of reality. Disbelief, even when confronted with proof, because we don't want to believe.

CHAPTER EIGHT

"You worry too much about the morality, man," the leader said earnestly to the thin one.

The fire was large and comforting, its reality lighting a room around them with defined and warm walls that penetrated outward to the tall trees on one side and the noisy water black below them on the other.

The others listened approvingly, nodding in agreement.

"My wife and kids deserve the best. And the group knows what's best. If I had made the decision, it would be different. But that ain't our responsibility."

"But an old man and a boy?"

"It's the principle of the thing, that's what counts. They're just ordinary people. The group is extraordinary. Different. We belong to that group because we are superior."

"Yeah, I agree, they're just sheep, all of them. But this seems more . . . more personal."

"It ain't your responsibility."

The finality of the statement was felt, and they began to bed down.

IS THERE A FUEL SHORTAGE?

There is such a plethora of unmitigated bull-roar surrounding the issue of the gas shortage that one can be totally confounded by listening, even with half an ear, as the controversy rages.

Let's try to examine some of the alleged facts. But, be careful, for I may be guilty of manipulating statistics just as facilely as all of the other participants in this arena. Perhaps my bias is to be grim and pessimistic as an argument for more rapid utilization of solar energy now.

I feel that the latter bias is the **result** of the facts, and not a facile rationalization to **cause** them. Certainly that was the order of their occurrence in my own approach to the problem. It has affected how the solar furnace invention was to be marketed. I would have preferred a more slow and cautious implementation, personally, but my interpretation of the fuel supply statistics — and more importantly the **consequences** involved — is scary. I sincerely hope my interpretation is wrong!

On May 27, 1975, Norman Lutkefedder of the Federal Energy Administration made some amazing projections. He stated that the United States' consumption of energy in 1985 would be the equivalent of 50 million barrels of oil per day. After ten years time, even a

A plethora of unmitigated bull-roar

Bias is grim and pessimistic

What are the consequences of the fuel crisis?

FEA makes amazing projections

vastly accelerated solar implementation program could not account for more than an equivalence of 1 million barrels of oil per day – a scant 2% contribution.

Demand substantially overstated

Two things about that statement are disturbing. One is the projection of a 50 million barrel per day utilization by 1985. We are presently using about 18 million barrels of oil energy equivalency per day, **down slightly from last year,** probably in response to higher prices and some conservation steps being taken. Even with an annual **10% cumulative** growth in energy demand the requirements in 1985 would only be 42 million barrels equivalency. A more reasonable projection would assume a continuing emphasis on better conservation methods and continuing price increases which would surely hold the growth to a 5% or less annual growth in demand for energy. This would result in a 1985 utilization level of 27 million barrels of oil equivalency per day, probably less.

Solar implementation understated

The second disturbing part of Mr. Lutktfedder's statement is the vastly accelerated solar implementation resulting in only a 1 million barrel equivalency per day by 1985. If a simultaneous program of retroinsulation and aggressive free enterprise unimpeded by hamstringing legislation are permitted, America's small businessmen can account for 4 million barrels of oil equivalency from solar energy by 1985, with only 50% utilization levels. That's almost 15% of our total energy supply from solar, and assumes no improvements to present technology. And you

68

may be sure that some amazing progress is going to be made in the next 10 years with clean, nonpolluting, safe solar energy.

The important question, however, is this: why would the Federal Energy Administration **downplay** potential solar energy contribution while simultaneously exaggerating the potential energy demand in an official statement?

In the raging debates of energy alternatives, coal is often proffered as the one fossil fuel in adequate supply. It is difficult to see **how**. Very little anthracite, but large deposits of bituminous and lignite are presently available. The latter two, however, will require energy-consuming processing before being utilized in today's energy market which must concern itself for the first time with the **consequences** of massive pollution.

The **logistics** of coal supply are staggering. Just **how** is that coal to be delivered to the users? Our rail system seems the only logical method for accomplishing delivery, yet preparing that system could require ten years and billions of dollars of investment. Equipment is outmoded or already junked, the rail companies are folding with systematic regularity, and the road-beds themselves would have to be revamped and upgraded to be used safely. Is the government going to do **that?** If so, one should conservatively double the time required to do it.

The natural gas supply is another arena. Here is the real problem area. We are presently using about 25-30

Why the exaggeration of demand, and disclaiming of solar?

Other alternatives: Coal not a panacea

Logistics of coal processing and delivery are staggering

69

Natural gas is running out rapidly

Severe shortages by 1980

Many homes on natural gas. Even if they are not supplied in 1980, huge industrial shut downs inevitable

trillion cubic feet of natural gas per year. The delivery system for this product has been added incrementally over a period of decades, and is already in place. Yet there is widespread agreement that tremendous shortages will occur during the next three years, with unsolvable shortages occurring by 1980. Four and a half years from now.

A chart published by the Federal Power Commission is shown below. Look at the year 1980. Assume that home heating alone takes 25% of that total demand indicated by the upper line. If that gas is supplied to homes, we are looking at unimaginable industrial consequences. During the height of the 1930's depression, the peak industrial shutdown nationally was 24% with a concomitant 24% unemployment. Take just a minute to study that graph. **You** be the future reader. What kind of industrial shutdown would you predict in 1980? Even if we don't supply natural gas to homes? If that little exercise doesn't scare the pants off you, you are, to put it bluntly, insensitive.

If this doesn't scare the pants off you, you're insensitive

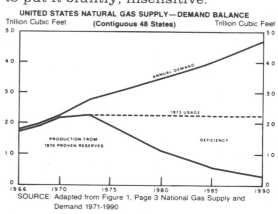

UNITED STATES NATURAL GAS SUPPLY—DEMAND BALANCE
(Contiguous 48 States)

SOURCE: Adapted from Figure 1, Page 3 National Gas Supply and Demand 1971-1990

FPC Bureau of Natural Gas, Washington, D.C. February 1972

Yet this year the National Academy of Sciences went on record as disputing that graph as **rosy-eyed** — that the proven reserves were only ½ as much as you see indicated there. If they are right, you should not just be frightened, you should be terrified. I don't think you'll panic. But I hope that when you examine the consequences — the immensity of the problem and its imminence — you'll begin to insulate your home and begin to take positive personal steps to begin conserving energy.

Ready to be confused just a bit? The Federal Trade Commission says that the Federal Power Commission graph **understates** the proven reserves by 25%. So if you are still harboring illusions that the crisis is imagined, go back once more and perform the same exercise in fortune-telling with the proven reserves at a 25% higher level. Even with the optimistic approach which sounds so comforting when stated as "25% more" we're in big trouble by 1981 instead of 1980. Big deal!

Let's assume that the Federal Power Commission Graph represents the governmental "moderate position." It is about halfway between the FTC and the National Academy of Sciences positions. What kinds of problems do we face in 1980?

I think that major dislocations of industry are inevitable. It is also inevitable that as these dislocations begin to make themselves felt, the reversal of service priorities **must** occur. Whereas, **TODAY,** the domestic user gets first shot at supplies, very soon the industrial user will get first shot. After all, in

National Academy of Sciences says these reserves are overstated

Federal Trade Commission says reserves understated

Federal Power Commission probably moderate position

Major dislocations of industry inevitable

Widespread unemployment inevitable

What is the alternative for the homeowner?

Many homeowners now paying more for fuel than for their mortgage

widespread industrial shutdown, the homeowner will be unemployed, so he won't be able to pay for the gas anyhow. Attempts to stabilize the economy, locally and federally, will have to take precedence. As a matter of fact, this kind of reordering of supply priorities is already taking place in various localized areas all over the country.

Let's look at consequences. What substitute may the homeowner find for natural gas? Coal? Probably not. Even more delivery problems are associated with residential use of coal than exist with industrial use. Developers have left the alley out of the subdivision for years. Which half of the rec room should be converted to coal bin? Who is going to market **that** to today's middle American? LP gas? Already in short supply. Costly beyond most people's means already. Electrical heating? Nope. Just aggravates the problem. It takes **fuel** to generate the additional electricity required – at low efficiencies. And to **heat** your home with electricity takes 20 to 30 times as much as you are presently using. Like adding 20 to 30 houses to the power network. "Peak loading" is the electrical utility's nightmare **today**. Conversion to electrical heating on a mass scale would be totally impossible. Fuel oil? Perhaps. But only with dramatically increased imports. And, again, the fuel oil user today, in many parts of the country, is paying much more than his mortgage payments just for heating. And that ain't gonna get better.

We are **presently** importing nearly 40% of our total energy · requirements. Energy is the basic denominator in **any** manufacturing equation....Just how far away is national bankruptcy at 40% level of energy imports? Our balance of payments problem gets so aggravated that we now are the arms supplier to the world, a change in national morality that is bound to backfire one day.

I am not an economist. I'm not sure that such a critter exists capable of relating to the enormity and complexity of today's macroeconomics, anyhow. My basic farmer logic, however, tells me that 40% energy imports is sheer national insanity. What kind of harvest are we sowing for our children to reap?

Just how are we to keep our families warm in 1980? Just 4½ years from now? You may be sure that the nuclear energy club is waiting in the wings to solve your problem **their** way, if you don't solve it some other way — **personally** — between now and then!

Energy is the basic denominator

40% imports is national insanity

How are we to keep our families warm?

CHAPTER NINE

The old man and the boy moved quickly, now, in the gray light before dawn.

They moved in that jarring, awkward lope that mountain men soon learn for coming down a slope. The valley opened below them, a gentler slope beckoning a few miles below.

The town could be sensed, rather than seen below the ridge that cut across the bottom of expanse ahead, lights and steam or smoke providing a brightening in the air.

"Will we go into town, old man?"

"Yes, boy. But slowly. We must stay ahead of the pursuers until then. Only in the heart of town may they catch us."

"What will they do with us then?"

"Nothing, boy. Don't be afraid."

They wound their way through the marshy bog of high country grass that surrounded a rivulet suddenly spread wide across the hillside, then down a steeper, rockier slope.

The shadow of the mountain behind them passed away, moving like a cloud upon the valley in front. Rushing at them rapidly. As it passed abruptly, they felt the warm red of sun on their heads and backs.

THE QUESTION OF RETROFIT

"Retrofit!" An interesting catchword indicative of the age we live in. Retroactive fit. Meaning that something can be added to an **existing** structure. Something compatible to that existing structure.

The importance of this catchword in solar heating applications is paramount, and, one would think, terribly obvious. Yet only one system has ever been developed with practical compatibility to existing homes. The backyard solar furnace. And it most assuredly was not developed by governmental funding.

No governmental funding has been devoted to retrofittable solar heating systems. Why?

There are from 65 million to 80 million existing homes in the United States today. From 300,000 to 1½ million **new** single-family homes are built each year.

Assuming that there is **some** validity to advocating solar heating in single-family homes as a way of conserving our rapidly depleting reserves of fossil fuels, one would expect that at least **fifty times** as much effort would be devoted to the research and development of retrofit systems as to the development of systems for new housing. Yet **no** gov-

A solar furnace must be compatible to existing homes

Only one retrofit system, the backyard solar furnace

No governmental interest in retrofit

50 times as much effort should be devoted to retrofit solar systems

ernmental funding was devoted to solving this problem.

90-95% of existing homes use forced-air heat

Second statistic. Depending upon which source you choose to believe, from 90 to 95% of all existing homes are heated with forced-air. Simple reason. On the average a forced-air system costs about ⅓ to ½ as much as a hot water system, regardless of the fuel: natural gas, propane or oil.

The retrofit system should be forced-air and should "plug in"

From this, we might, without any fantastic flight of deductive logic, arrive at a key constraint for retrofit systems: they should interface easily with forced-air heating systems, since that is the equipment already existing in 90 to 95% of the homes. (If I'm going too fast, stop me!) In other words, the ideal retrofit system should be designed to "plug in" to the existing home.

Rooftop installations are out of the question

Another important consideration is the keeping of expense and remodeling to a minimum. A solar collector designed specifically for heating should be oriented at about a 60° angle facing south. (Technically, at the latitude plus 20°, so that a home located at 40° North Latitude would have a 60° angle collector, at 35° North Latitude a 55° collector angle and so on). Although absolutely no statistics are available, I will go on record as hazarding a guess that no more than ten existing houses in the United States have a 60° pitch roof facing south!

All governmentally funded projects use rooftop collectors

Yet all of the governmentally funded projects designed to make solar heating practical utilize roof-mounted collectors. One can conclude that it would be possible to retrofit ten houses in the U.S. with rooftop collectors without replacing the roof. A bothersome procedure that, replacing a roof.

Further, no one has perfected (yet) a neat way of adding a 5000 gallon storage tank to your existing basement. The builders of the past inconsiderately forgot to provide a door that large. Second alternative would be to bury the monstrosity in the backyard. Simple, right? Few problems, however. Typical backyards, without alleys in many cases and with privacy fences very usual, present real problems to the access of a huge Caterpillar tractor, a front-end loader and two dump trucks. Your wife may object to the desecration and destruction such a project would involve, as well. Imagine that going on in your **present** back yard. Kind of a personal strip mine, right there in your tulip patch!

Almost impossible to retrofit a 5000 gallon water tank

Now remember back to all of those fantastically practical systems you've been reading about for the past two or three years – most of them built with **your** money. Huge collectors on the roof, storage tank in the basement. That's been the governmental approach to developing practical solar energy utilization. Notice also that they always employed **water**. Remember any forced-air systems being built by the government?

Governmental practicality— use water tanks and water systems

The **real** question is, with all of the academic and giant corporation input to the governmental funding of solar research, why hasn't at least **one** retrofit system been built – especially when the need for retrofit is 50 times as great?

Why no government development of retrofit systems?

Scientists stupid and/or incompetent otherwise, oversight deliberate

The **least damning** analysis one can make is that scientists receiving governmental grants are **stupid and/or incompetent**. The alternative conclusion has to be that the "oversight" was **deliberate**.

CHAPTER TEN

The thin one jumped as the scout materialized in front of him in the darkness.

"You wondered, too?" the scout asked him.

"Yeah."

They climbed on in the ever brightening gloom.

"They may need help—the old one and the boy."

"Yeah."

Together, they altered their course, heading more westward, picking their way to the top of the ridge.

They started down the valley opening before them. As the sun rose about an hour later, they saw the pair walking ahead of them about a half mile.

"Halloo!" shouted the thin man, before the scout could stop him. He was surprised when the pair looked back and began to run, straight toward the town.

The scout began to run, as well, and he shouted, hands cupped to a megaphone shape around his mouth, "We got your note!"

CONSUMER PROTECTION AND "CERTIFICATION"

One of the most interesting departures from logic and sound reasoning is the paradox of some consumer advocates.

The approach in solar applications is something like this: the **government** should certify privately manufactured solar heating systems and set standards for systems so that the consumer may be protected.

That has to be one of the most direct examples of internal contradiction ever forwarded.

Taking a government which has demonstrated in recent history a consistent and unswerving total ineptitude at waging war (the traditional forte of governments, **n'est ce pas?**) bumbling ineptitude in the running of a simple postal system, outrageous inadequacy in providing a decent social security insurance program, staggering incompetence in formulating strong and sound energy and environmental policies, sardonic and unremitting humor in the regulation of radio and television, the airlines, interstate commerce, breathtaking creativity in developing a totally (in principle) incomprehensible set of taxation procedures — **and then to have the unmitigated gall** to **encourage** that government to establish a consumer protection bureaucracy which will, because the goal is

Complete departure from logic

"Government should certify solar heating systems"

A government beset with bumbling ineptitude, staggering incompetence in regulating should regulate. Where's the logic?

83

laudable, function efficiently, usefully and fairly is unbelievable!

Don Quixote is not dead!

Don Quixote is certainly not dead. Naivete of the highest order yet achieved in history has been formulated.

Government dominated by big business already

Consumer protection groups must of necessity **not** be associated with the government. Government **already** shows itself to be dominated by the giant corporate interests to the detriment of the populace.

Private public interest groups should watch with a cynical eye

Typical industry self-regulation should continue to be encouraged. But it should be watched with a cynical eye by **private** public interest groups. Are the self-imposed standards of conduct (the typical "code of ethics") being rigorously observed, or is mere lip service being rendered?

The media can serve as a watchdog

The media and public interest groups are a powerful and positive watchdog, and their regulation **works** by hitting the businessman where it's sure to get his attention – in his sales. Despite the possibility of business abuse, the small businessman dominated regulation organizations – the Better Business Bureaus and Chambers of Commerce – do perform and have performed significant and useful functions.

Self regulation can work

Trade organizations, such as the Realtors, who live under a self-imposed set of rules for conduct **do function** in self-policing ways.

Flim flam can't survive public exposure

Flim flam just can't survive the light of exposure. Consumer protection advisory services, such as Virginia Knauer's office, come closer to hitting the proper utilization of government in consumer protection areas—by promulgating **educational** materials and insuring that **media coverage** is provided. Protecting the consumer **after** the barn door is open is necessary, but the real answer is in educating him to **recognize** the flim flam.

Now we come to "certification". Late last year, along with a few others, I voiced loud criticism of the governmental effort to establish the National Bureau of Standards as a "certifying agency" to regulate the new solar industry. Evidently the outcry was sufficient for at least some reconsideration of the advisability of establishing NBS **for the first time** in a regulatory position with private business.

Certification by the National Bureau of Standards

At that oft-referred-to SEIA-FEA-ERDA Expo 75 conference in Washington, D.C., some of the officials from other agencies indicated some dissatisfaction with the NBS approach. That should have been that. Certainly, the NBS attempt to enter the field of regulation had far exceeded the directive to it from Congress. Another incident occurred during that conference, however, which fed my incipient paranoia.

NBS approach in some disfavor

International Solarthermics Corporation held a small press conference that first day, to announce that practical solar heating was here, **now**. That 56 separate small businessmen—manufacturers, distributors and dealers—operating from coast to coast were now selling and installing retrofittable, practical solar furnaces. That the furnaces were not only installed with structural and component warranties ranging from 5 to 20 years, **but were**

Press conference to announce 56 small businesses selling from coast to coast in new solar industry

also being sold with a performance warranty standing behind the actual heating performance of the systems. That there would be over 1000 companies and corporations involved in this new industry by the end of the year. That at least 3500 additional solar heated homes would be on stream by the end of the year. (Governmental funding over the past years has accounted for a grand total of less than 100).

Private industry moving faster than FEA accelerated projections

One must suppose that in response to two stimuli — the general dissatisfaction with the NBS Standards and the announcement that private industry was already moving toward practical implementation of solar energy far in excess of his publicly stated "accelerated implementation" projections — Norman Lutkefedder, of the Federal Energy Administration, **that evening** hurriedly, by phone, arranged a **closed-door** meeting of ERDA, FEA, NASA, HUD and NBS officials with the private industry standards setting groups of ASTM and ANCI. As an on-the-spot self-appointed spokesman for the **real world** participants in the solar manufacturing and distribution areas, I asked Norm Lutkefedder about attending these meetings. The crisp answer was that the meetings were **private**. That two meetings were to be held in star chambers. That minutes of the meeting would **not** be available. After those (strategy?) meetings, why, **then** public discussion would be invited. In what open and participatory ways do the **representatives of the people, paid by the people**, work? What does FEA have to hide?

Closed door meetings to be held with no minutes taken

I've been spoiled by the Colorado Sunshine Law, where officials are not permitted to have closed door meetings. The title of the law, alone, is enough to make it appealing to me, but the principle behind it is one that in these post-Watergate times the Federal Government had better get prepared for — it's comin' guys, and you're going to have to start working out in the open where we can see **you**, too.

In fairness to the FEA, which, despite its "newness", is another burgeoning bureaucracy, I must make some other comments.

There is a **crispness** in FEA not found in other energy agencies. They are well-briefed and aggressive. They are **action-oriented**, in my opinion. This is very refreshing in government, but also quite frightening. Their commitment to the implementation of nuclear energy is, unfortunately, quite well seen. Mr. Zarb recently advocated 20% energy supply from nuclear by 1980. I sincerely hope that this young agency, which by its dynamic and fresh approach might do some very positive and useful things to solve our energy crisis, is not being totally and callously manipulated in their actions by the pressures from the giant nuclear corporations. I am not surprised that their approach to solutions is via nuclear — they are after all, obtaining much of their "factual information" and statistics from the ERDA group which is largely staffed by ex-AEC people.

This problem of information gaining is complicated by yet another factor. The small businessman is, for the most part,

Federal Sunshine Law needed. Our representatives and paid employees should work out in the open— where we can see them

FEA crisp and action-oriented; this is refreshing and frightening

**Small
businessman
works too
hard, too
many hours
to be a good
communicator**

not a good, **participating** citizen. Why? He is required, of necessity, to work 14 to 18 hours a day, often seven days a week. His communication with the world outside his business, therefore, is severly limited and likely to be confined to reacting to events **after** they have occurred, as opposed to participating and supplying input leading to the formulation of a governmental decision impacting on his life.

**Government
should
aggressively
seek the
grass roots,
small
business
input**

He often cannot afford the time or expense of traveling to Washington to communicate with the policy makers. He also can't afford a lobbyist. Communication should be a two-way street. The policy makers should recognize this problem of the small businessman and **seek** his input prior to making a decision affecting that businessman. A corps of ombudsmen, capable of traveling the United States, should go out to the people and seek their input.

Laudable laws have a way of backfiring and hurting the small businessman to the advantage of the giant corporation — perhaps accidentally? The best recent example of this is the OSHA legislation. Small, ten-man machine shops, for example, which have functioned safely for thirty or forty years with no more extreme injury than one requiring a few stitches, are suddenly being forced out of business by unrealistic sanctions which require the installation of safety equipment which cannot be borne economically by the small entrepreneur. Surely the intent of OSHA was **not** to close down the small business, but rather to improve the occupational safety of the workers. Go talk to the workers in a small shop and

**Surely the
intent of
OSHA was
not to harm
small
business, but
that has
been the
effect**

see for yourself if the "boss" is unconcerned about their health and safety. **He's** working under the very same conditions they are! Notice that "boss" in the small business is a totally different concept than "manager" in a huge factory.

A study of small business failures **due** to OSHA-imposed regulations is long overdue. How much of the present unemployment is due to this single source? Surely the OSHA investigators could proceed with more commonsense and temperance? One can easily imagine the small farmer being phased out as well, with this mechanism of government. Were the abuses to life and safety in the **small** businesses or the ones in the **big** businesses the abuses that the OSHA legislation was designed to remedy?

Let's not put "consumer protection" in that arena of unfeeling interpretation of ambiguous "rules" by overzealous, self-important investigators, particularly if, accidentally or otherwise, the harm accrues yet another time to small business to the advantage of giant corporations. The consequences could be extreme.

How many small business failures due to OSHA?

Consumer protection doesn't come from unfeeling rigid interpretation of "rules" by overzealous officials

CHAPTER ELEVEN

The leader woke with a jump in his leg. The sun was shining in his face. Hummingbird squealed past like an oversized bee, disappearing in the trees.

He looked about the camp. Damn. The scout and the thin one were gone. Should have guessed it would happen when he saw the thin one pick up the piece of paper surreptitiously behind his back as they were moving out yesterday. This would look bad on the report. But it wasn't his fault. He hadn't made the decision of just who would go. He'd have to plant that seed of accusation-suspicion in the paperwork. They'd have to move slower, now, without the scout. Damn and double damn.

Angrily he went about the camp, kicking the others awake.

The camp jay watched them eat quickly, and waited patiently, basking in the sunshine, ruffling its feathers occasionally.

MONOPOLIZATION OF THE SUN

Thus far we have seen a very hard to understand governmental approach to solar energy which is either incompetent or devious. Either no one saw the necessity for implementation at the single family residence level of solar heating, or there was a direct attempt to portray solar heating as ungainly, impractical and expensive, by selecting for demonstration and research **only** the most impractical of the impractical. It is hard not to believe, after seeing the tremendous expertise involved in the field of solar applications, that any conclusion can be arrived at but a conspiracy to hinder the implementation of solar energy utilization.

For the last time I return to that infamous "trade show", EXPO 75, for the **new thrust** of the government, hand in hand with the giant energy corporations. Was it a trade show, with manufacturers displaying their wares and trading technology? Nope. It was a forum for ERDA and FEA. They dominated the program for three days, and they showed where solar was **really** going.

No slides showing working systems. No proud homeowners posing by recently purchased systems. No unveiling of significant and practical advancements by the real world manu-

Governmental approach to solar either incompetent or devious

A conspiracy to hinder implementation of solar utilization

The new thrust of government and industry exposed

93

facturers operating out there in the marketplace.

Madison Avenue glibness

Instead, the future was unveiled, with the public relations Madison Avenue glibness prevalent in today's government, a barely contained excitement in the air. A hale camaraderie expressed itself—"see, fellas, we really are working for solar implementation."

The future the energy giants envision: frightening

It was a frightening future they unveiled. The kind of future the energy giants desire, the future they envision for you and me. Partake.

Ocean-thermal Conversion— Giant solar islands

Item. Ocean-Thermal Conversion. This is a scheme where giant concrete islands would be floated in the Pacific Ocean, drawing cold water from the depths, warmer water from near the surface and using the temperature differential to power huge turbine electrical generators. Complete with beautifully airbrushed artist's conceptions showing man's technology at work in the future. Environmental damage possibilities casually dismissed as negligible, since they would shoot the used cold water back toward the depths. Creatures inhabiting the millions of gallons of water rushing through the device would evidently not be harmed. At least no allusion was made to that aspect.

Solar farms— 40 acres of mirrors

Item. Solar Farms. Where twenty to forty **acres** of mirrors are focused to a tower to generate steam and hence electricity.

Item. Bio-conversion techniques. Where specialized plant crops could be grown under accelerated conditions on a mere 200,000,000 acres of ground (lying fallow somewhere, no doubt) to be harvested and burned as direct fuel, or processed to form other fuels. Environmental considerations: "Minimal". The sophism required to treat a plant as stored solar energy which might be used as fuel while others are predicting world food shortages due to unavailability of tillable land is adept. One is reminded of similar techniques of using language to which we have all been treated in the recent past.

Item. Photovoltaic conversion of solar energy. Where dissimilar materials bonded or "grown" together produce electricity. The familiar solar cell. Practicality is just around the corner here. "With mass production using silicon, the most abundant element in the earth's surface, there's no environmental impact here". No discussion of the rare earths used. Cost is tossed off lightly. Present price of $1.7 million per all-electric single family residence can be brought down ten times, maybe twenty times.

(Not mentioned directly, probably because NASA was not an **active** participant, was the satellite solar collector which microwaves the energy to earth.)

Item. Giant wind energy generators feeding electricity into the national electrical grid.

Well, gentle reader, is it beginning to make sense? ERDA is going to support high technology **centralized** solar energy utilization. Conspicuous by its absence was the emphasis on **private ownership** of solar heating systems. Real lack of enthusiasm for that concept on the part of ERDA.

Bio-conversion: 200,000,000 acres turned to fuel instead of food

Photovoltaic solar cells —cost tossed off lightly

Windmills generating electricity for the utilities

Centralized control of energy production is the goal

**Decentralize
solar
ownership
gives
individual
independence**

**Centralization
production
permits
charging
you for
sunshine**

**Strengthen
the energy
suppliers'
stranglehold**

You may remark, "well, it's better to have centralized solar than nuclear generation plants with the concomitant hazards." Agreed. But examine the issue of centralized versus decentralized solar implementation, and see what the long-term implications and consequences are. Decentralized solar applications mean that the ordinary man owns his own utility. Right on his property. Self-sufficiency and energy independence, or at least the first step in that direction. How does the giant energy supplier control this man?

Contrast that picture with centralization of solar collection. Even though enough energy falls on your property to handle your energy needs, you allow them to produce the energy elsewhere, transport it to your home and charge you for the use of sunshine.

If you prefer to condemn not just yourself but your children and children's children to paying throughout their lifetimes for sunshine, to strengthening the giant energy cartels' economic stranglehold on governments and nations throughout the world, then centralized solar collection is the way to insure it.

Six out of the seven operational branches of ERDA are committed to that goal — read the interim organizational chart and see for yourself.

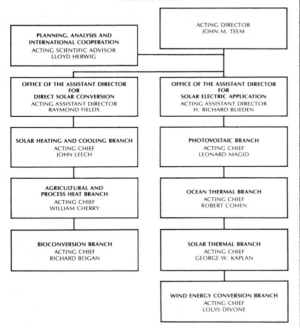

ENERGY RESEARCH AND
DEVELOPMENT
ADMINISTRATION DIVISION
OF SOLAR ENERGY INTERIM
ORGANIZATION

ACTING DIRECTOR
JOHN M. TEEM

PLANNING, ANALYSIS AND
INTERNATIONAL COOPERATION
ACTING SCIENTIFIC ADVISOR
LLOYD HERWIG

OFFICE OF THE ASSISTANT DIRECTOR
FOR
DIRECT SOLAR CONVERSION
ACTING ASSISTANT DIRECTOR
RAYMOND FIELDS

OFFICE OF THE ASSISTANT DIRECTOR
FOR
SOLAR ELECTRIC APPLICATION
ACTING ASSISTANT DIRECTOR
H. RICHARD BLIEDEN

SOLAR HEATING AND COOLING BRANCH
ACTING CHIEF
JOHN LEECH

PHOTOVOLTAIC BRANCH
ACTING CHIEF
LEONARD MAGID

AGRICULTURAL AND
PROCESS HEAT BRANCH
ACTING CHIEF
WILLIAM CHERRY

OCEAN THERMAL BRANCH
ACTING CHIEF
ROBERT COHEN

BIOCONVERSION BRANCH
ACTING CHIEF
RICHARD BOGAN

SOLAR THERMAL BRANCH
ACTING CHIEF
GEORGE W. KAPLAN

WIND ENERGY CONVERSION BRANCH
ACTING CHIEF
LOUIS DIVONE

The giant energy suppliers, however,
never put all of their eggs in one bas-
ket. Suppose that the centralization
thrust doesn't work?

97

CHAPTER TWELVE

"What did he say, boy?" panted the old man as they jolted down the hillside.

"Something about the note, I think."

"Ah", smiled the old man, stopping abruptly, "They did get a chance to read it, some of them."

The boy ran on a few paces before stopping to see the old man bending to find a sitting place behind him.

"What if it's a trick, old man?"

"We have nothing to fear, boy."

And the boy, puzzled, panting, believed the old man, and dropped to the ground beside him. They watched the scout and the thin one stop running behind them, walking now to approach them.

"We have no weapons," shouted the thin one, as they got closer.

"Why, they're afraid of us, old one!", said the boy, in amazement.

"Yes, boy, that is the secret, isn't it?"

The scout and the thin one approached, uneasily, and at the invitation of the old man, sat on the ground beside them.

They talked.

THE LEASING LIE

By now it should be obvious. **Control** of the energy source is **always** the goal. But how does one **control** the **sun?** Some of the ways have been explored. But failing all of those, a last ditch attempt is still available, and the groundwork for that is being laid today.

The energy suppliers watch with envy the telephone company approach. The "leasing" or, more properly, the "renting" of systems. That the latter is being contemplated as opposed to the former is clear from some of the language being used.

The campaign began some time ago and has already had significant impact on some political agencies. As usual, there is the "big lie" — one which rolls off the tongue easily, is eminently quotable, and is often "bought" by the news reporter trying to do an honest job of presenting a story with the broad perspective.

Here's how it goes: "Solar heating equipment is very expensive. Because of high initial cost, **only through leasing** will the equipment be available to the average homeowner."

Sounds reasonable, doesn't it? Commendable intention: to provide for the poor bloke who couldn't afford it otherwise. Sorry. You've been suckered yet another time, friend.

How does one control the sun? Easy

Energy suppliers envy the telephone company

The "big lie" rolls off the tongue easily

"Solar equipment costly— only through leasing can ordinary man afford it"

Bankers will loan on solar equipment today. No money down

Let's look at the **facts**. Bankers from all parts of the United States are ready to loan the homeowner the money necessary to purchase and install a solar furnace. Second mortgage, home improvement loan. In some cases, with a preferential "energy conservation" interest rate. No down payment if the homeowner has lived in the house from six months to a year, depending upon the lending institution. The interest is less than that paid for an auto loan from the same place. So the homeowner buys a system, with **no** down payment, pays principal and interest for 5 to 7 years and ends up **owning** the system.

Lease costs a deposit plus higher monthly payments

Contrast **that** with the big lie **lease**. A typical lease will not include installation costs. The homeowner foots that bill. He will also pay a **deposit** equal to the first and last months' payments — on a 5 year lease, a deposit of six month's payments. He must have as good, if not better, credit to qualify for the lease as he would have to have to purchase. Finally, his **payments** will be higher, because they will include not just principal and interest as in the case of a purchase, but also a profit for the leasing agency. He may or may not be allowed the privilege at the end of the lease period to purchase the equipment. So the "lease" which sounded so good when stated earlier, turns out **not** to help the average homeowner, but to stick it to him.

Even "renting" costs more

Now, if the "renting" approach were followed, the homeowner would **still** pay for the installation charges initially. Perhaps, if the "renting" corporations were magnanimous, his payments would be less than his principal plus interest costs in a purchase. But the "initial high costs" rationale is in either case a smokescreen intended to divert attention from the real purpose, i.e. **control**.

The energy suppliers would like to tie up this particular area. I attended a workshop sponsored by the National Academy of Sciences and Engineering held in Tempe, Arizona on December 11, 1974. A spokesman from Southern California Gas Company made a strong pitch to the manufacturers present that they should sell their equipment directly to the utility. The content was simple and aimed directly at the manufacturers' interests along these lines: If you sell to a utility, there's no warranty work involved, no service organization, no dealers, no competition to speak of, so you come out ahead. Clearly, there's much more profit for you if the normal distributor-dealer organizations are not required. No credit or accounts receivable problems, either.

Selling the equipment has no direct appeal to a lot of the energy suppliers, because it is a one-time sale, and control is lost.

The energy suppliers toss yet another red herring out to make leasing the "only solution". Many of the governmental personnel have bought the concept that financing of solar heating systems will be impossible without governmentally backed loans. Earlier in this chapter I said that bankers were ready, willing and able to loan on solar furnaces. Why?

Telephone company charges you an installation fee

Energy suppliers courting the solar equipment manufacturers

Selling equipment has no appeal because control is lost

103

Bankers are personally aware of hardships of fuel crises

Harder to qualify for new homes due to inflation and fuel costs

Solar furnace is an investment in real property, qualifies for real property treatment

The banker is aware of the energy crisis in a very personal way. Many of his clients come from rural or suburban areas where natural gas is not being utilized. He has seen these clients have problems when their fuel bill outstripped their mortgage payments. He can see an increasing foreclosure rate looming in the future as this problem gets unmanageable. He has in part been responsible for the slowing in the construction industry from the other side. He has increased his earnings requirements for qualification for a mortgage to account for both inflation and increased fuel bills. The solar furnace, if it is a tested, warranteed, safe product, produced by a reputable manufacturer, represents a real estate investment. The home improvement loan made to purchase it is secured by real property, which represents excellent collateral. He doesn't talk about "amortizing the unit's cost against fuel savings", because he knows better. Real property tends to appreciate, year by year, and it has done so at an average rate of 4% per year for the last hundred years, at a higher rate in the past few years. Unlike an auto which depreciates, his collateral will be increasing in value. And it represents increasing equity to his customer—equity he gets back when he sells the home, or equity which is reflected in his net worth if he keeps the house. Finally, the cost, installed, of a solar furnace in the range from $3,000 to $6,000 fits nicely within the typical range of home improvement loans.

Therefore, the local banker does not need governmental incentives to make a loan on solar furnaces.

Partial or total subsidies to give the private homeowner incentive to install solar equipment are great. I like to view such subsidies as "reverse depletion allowances" rewarding the individual who voluntarily stops depleting our valuable fossil resources. If oil companies got depletion allowances, the homeowner deserves an allowance for **not** depleting.

Government should institute "reverse depletion allowances" to reward homeowner who cuts utilization of fossil fuels

CHAPTER THIRTEEN

The group of men crested the ridge sweating, the leader angrily goading them on. They stopped for a moment as the panorama unfolded, because no man could do otherwise. It was so expansive. One of the group saw the party of four sitting on the ground below them.

"Traitors!" muttered the leader as he recognized the scout and the thin one.

The group below saw them then, and stood up, the boy dusting the back of the old man. They began walking, leisurely, toward the unseen town over the ridge.

"They're heading for town. We'll catch them there." the leader said thankfully. It was almost over, this assignment. They swarmed down the hillside behind the departing foursome, running, jogging to close the gap.

REGURGITATION

The Players in the arena:
  Giant industries
  Small industries
  College professors
  The Federal Energy Administration
  The Energy Research and
    Development Administration
  The National Bureau of Standards
  The National Aeronautics
    and Space Administration
  Agency of Housing and
    Urban Development
  (Deceased, but not dead)
    Atomic Energy Commission
  Advisor to the President,
    Consumer Affairs
  Nader's Raiders
  Federal Power Commission
  Federal Trade Commission
  National Academy of Sciences
  National Science Foundation
  Assorted Bit Players

The manipulators, the manipulatees

In general, one may categorize the first of that lengthy list as the manipulator, the rest as manipulatees. Several of the manipulatees, in turn, are either willing or unwilling second order manipulators, functioning in one of three ways: 1) creating direct impediments to the introduction of solar energy utilization, 2) by creating chaotic confusing communications which serve as red herrings distracting from the central issue: the reality of the impending

Create impediments, confuse issues, show solar to be impractical

energy crisis, 3) Graphically demonstrating the impracticality of single family utilization of solar energy while arguing forcefully for centralized production of electricity from solar sources.

Is there a solar conspiracy?

The issues that cloud the battleground:
Is there really an energy crisis?
Are we really running out of fossil fuels as quickly as 1980?
Is nuclear power safe?
Is there any alternative to nuclear?
Can solar energy be a solution to the energy crisis?
Is it better to centralize solar production and sell it to people, or is it better to let the ordinary man own his own means for solar production?
Can the consumer be protected in the area of solar devices?
Only the government can protect the consumer?
Is there a solar conspiracy?

I may be stroking you with this book

As you can see, the game is clouded by many charges and countercharges from the players in that arena. Hell, I may be stroking you with this book! But I think that if one creates a simple philosophical position and measures **all** of the statements in this new field against that position, he **can** come to a logical conclusion.

Assume a philosophical position

Man's inalienable right to use the sun

Having taken so many liberties already, therefore, I take one more—the formulating of what I think should be the philosophical position.

"It is man's inalienable right to utilize the energy from the sun. He should be able to use that energy without restriction. If he desires to own equipment privately to produce useful energy for his family, as opposed to paying other men for that energy, also gained from the sun, that is his **right**. Whether giant corporations or governments own the production means the little man gets screwed. In the former case, the little man **pays** for the use of sunshine, in the latter, he is **taxed** for the use of sunshine. Either is a perversion of his rights, no matter how much rhetoric surrounds the actions."

Assume for the moment that what I now say is true: solar energy can be utilized to handle **all** of man's energy needs. The technology to do that exists **today**, not in some nebulous nether world of the future. It is less costly in the long run to utilize solar energy than any other source of energy.

Solar energy sufficient to handle all energy requirements

If those statements are true, measure all of the proposals with which you are bombarded daily. Why in God's name would we continue to trifle with the future livelihood of our children by the proliferation of potential bombs all over the country masquerading as "useful utilization of atomic energy"? Why do we continue to foul our atmosphere by burning fuels which could serve as the warehouse of resources for untold generations to come — not as fuels, but as basic building blocks of useful plastics, textiles, etc.?

Why trifle with our children's future by using nuclear?

There is a real and vital **war** going on right now. None the less evil just because the graphic pictures of innocents bloodied on the sidelines are not being relayed to your living room. We've got between four and five years to gear up and convert to solar utilization. And

This is a real and vital war

111

that conversion is certainly no worse than the alternative conversions — one of which you will be required to make during that same time span: you will have to either convert to electrical heating, the energy being supplied by behemoth belching coal plants or by the insidious nuclear plants, or, you will have to convert to coal or fuel oil. None of those conversions is cheap. I'm sorry. But you've got no choice. You can't continue to use natural gas.

At least take a step toward individual self-sufficiency

It seems simple logic, however, that if the conversion to something else **has** to occur at your home in the next four to five years, that it might as well be a **planned** conversion this time. One which takes into account the **consequences**. If you convert to electricity, fuel oil or coal, you have made a postponement of the inevitable. Nothing else. If you convert to solar utilization, although it is not a panacea, you have at least taken a real and concrete step towards your own, individual, personal self-sufficiency.

CHAPTER FOURTEEN

The old man held the barbwire apart, letting the other three crawl through. The thin one turned and held the strands apart for him. They struggled up the embankment to the pavement leading into the town from the south. No cars this time of morning. Quiet morning, they walked more easily now as they approached the small town.

"Where will we wait, old one?" asked the boy.

"Town square, I suppose. 'Cross from the hardware store."

They found the old mortar and rock benches and sat in the warming sun rising higher above the buildings now. The pursuers would be here soon.

BUILDING A BATTLE PLAN

What can you, personally, do?

One thing is conservation. But conservation just doesn't have much romance. Kind of like the grasshopper and the ants, and we all know in our innermost thoughts that all mankind is comprised of grasshoppers.

No romance in conservation

Perhaps I have convinced you that solar energy is the only solution to the dilemma facing us all right now. But you aren't feeling impelled to rush right out and put a solar furnace on your home. I don't blame you. Maybe it is worth waiting a year or so to do something, even if it does cost you a lot more.

Not in a hurry to convert to solar?

Well, there is something you can do which won't put you in the poorhouse, and will at least give you the satisfaction of knowing you're acting on the problem, **personally**.

No, I'm not going to recommend that you write to your Congressperson. There's a few of you who will, because you have in the past. The rest of you won't, anyhow. I think I understand your reasoning for not doing so, and perhaps I subscribe to that feeling, too.

Won't recommend you write your Congresspeople

Play a little game of imagination with me, though, will you? Suppose that it is 1976, or '77 or '78. Suppose, that for whatever reason, the natural gas to

Play a game in imagination

How do you provide for your family?

your home is shut off. Do you know where to shut off the water to your home so that the pipes won't freeze and cover your new shag carpeting to a depth of four inches of ice? Can you figure out a way to get water for your family with the water shut off? Have you got some way of keeping at least one room livable? (Don't count on wood burning. Try coal, as a better choice). If the gas is shut off for even a week or two, you can bet that everyone will be trying to heat with electrical space heaters, so figure on getting a couple of kerosene or battery lamps or candles to carry you when the lights go off. Have you got a way to cook without either gas or electricity? Probably should have.

Prepare your castle for a seige

Now, with that picture firmly in your mind, with **your** family being subjected to the problem, think a bit about conservation in a very practical way. For example, how difficult would it be to spend one weekend making your house a better heat holder? You should put 18 inches of fiberglass batts or the equivalent amount of insulation in your attic. The stuff's fairly cheap, and it's really easily put into the attic, outside a bit of sweating and itching. Check out the insulation in the sidewalls of your house. If it is practical, bring the quality of that up, too. Add storm windows and doors to your house. Can't afford them? Well, there's a poor man's way to accomplish the same thing. Read about it in Appendix A to this book. Probably the easiest thing, and the cheapest, is to cut the amount of cold air leaking into your house. A few tubes of caulking and some weather-stripping can really do wonders for cutting the heat loss from your home. Remember that exercise in imagination, and decide whether or not it might not be easier to keep your family warm with less of a heat source if your home is really well insulated. You'll get a double bonus out of being just a little scared – enough to **do** something about your house. You'll be able to cut your present heating bill in **half**. Next winter you can compare heating bills with your neighbors. Yours will be worth talking about, because maybe that will move your neighbors off their duffs for one weekend to do the same thing!

At least one bonus—you cut your fuel bill in half

If everyone in the U.S. did just that, we could buy some time. Not much, but some.

Second thing **you** can do. I'll not ask you to become involved in the public action groups which are fighting the introduction of thousands of atomic plants all over the country. I know how busy you are. But, at least, give your neighbor who is doing that thing your support. Pat him on the back once in a while, so he'll keep doing it. He really isn't a screwball. He just has a more vivid imagination than you do, and he's conjuring up even worse futures in his mind.

Pat your neighbor on the back

Third thing you can do. Now I know that you don't like to write letters. Neither do I. But you probably know someone, in your family, or among your friends. Con **him** into writing a letter! Tell him you'll sign it when it's done, as an affirmation of your support. That'll make him feel good, and he'll

Find someone to write letters for you

117

probably do it. Let him write to his Congresspeople. Never hurts. But the **important** ones to write are these:

> Frank Zarb
> Federal Energy Administration
> New Post Office Building
> 12th Street & Pennsylvania Ave.
> Washington, D.C. 20461

> H. Richard Blieden
> Energy Research And
>   Development Administration
> Washington, D.C. 20545

> Robert D. Dikkers
> National Bureau of Standards
> Building 266, Room B146
> Washington, D.C. 20234

They're the ones running things anyhow.

Have the letterwriter tell them that you people, as a group, don't cotton to paying for sunshine. That maybe the risks of nuclear are a bit grim for us to be willfully implementing it wholesale. That you'd just as soon be allowed to own your own solar equipment. Tell them that you have already taken the steps necessary to cut your residential use of heating energy in half.

Help your neighbors

The final thing I'll lay on you is an attempt to make you feel guilty enough to do something extraordinary. Know Mr. and Mrs. Jones, the old-age pensioners on your block? The ones who worked hard and long to pay off the mortgage on their house before retirement so they could survive on the social security and old age pensions that we as a society dole out to them so gener-

118

ously? They've got a real problem **today**. Their fuel bill is already, in many cases, more than that mortgage they worked so hard to pay off. Spend one weekend helping **them** get their house properly insulated. Old man Jones may not be up to crawling around in the attic putting in the fiberglass batts. Same with the caulking and weather stripping. Cutting their fuel bill in half will be an unbelievably important Godsend to them. They're your **neighbors**. Your dad or granddad knew what that meant in the early days, when it wasn't considered gauche to help a neighbor, with a barnraising or a community project to help with the plowing and seeding when a broken leg or sickness made it impossible for the neighbor to do it himself. Ive' probably gone too far with this suggestion. Sorry. Tell that lunkin' teenage son of yours to pull himself away from the TV set, and maybe pay him a couple of bucks an hour to do that for you. No interference with your schedule that way, and the result will be the same.

I guess by now I've bared my naivete to the world. I confess freely to nurturing a secret desire to see a return to earlier values of right and wrong, to the forgotten feelings of responsibility and obligations to one's neighbors that made America great. I bought the brainwashing of my youth, that self-sufficiency and independence are worth fighting for. I'm not a joiner, because I bought the concept of self-sufficiency so hard that I have trouble accepting someone else's idea of what that independence consists of. So I'll not ask you to join my cause.

Self-sufficiency and independence worth fighting for?

Make a conscious decision

Please think about it, however, and make a conscious decision of what **your** cause is. Now. Fortunately you are not powerless in this arena, as you are in a number of others. I understand your apathy in those other arenas, because I can't see the way to be effective in accomplishing anything except empty talk which goes nowhere, either. But in **this** arena, you **can** do something. Something concrete for you and your family. Perhaps for your neighbors, too.

But, hey. Do it **now**. Time's running out. Fast.

CHAPTER FIFTEEN

The leader stared at the foursome balefully. Sweat and dust on his forehead, rivulets of black running down his chin. Angry. The others circled the benches, preventing escape.

"You broke the law, old one."

"Whose law?" the old man smiled at him and at the gathering group of townspeople.

"I used the sun, nothing else. You are using the sun right now, as are we all. It warms the earth and the sky and moves the winds about us. How would I stop that use for a mere law?"

A murmur went through the bystanders, and the leader heard.

The thin one stood, now, and shaking his head, moved through the crowd. The scout stood slowly and moved behind him. The rest of the group moved away, leaving the leader to confront the old man and the boy alone. The old man smiled again and said quietly, "It's over. Time for breakfast, now." The leader smiled, at last, "You're right, old one."

The boy, watching, understood, finally, and rose to follow the men to the cafe.

THE SOLAR CONSPIRACY

The conspiracy? No, I haven't forgotten it. Even though the past few pages may have made you think so. You see, I've been goading you and me to **move**. Conspiracies thrive only in darkness, and only when **no one** raises a hand in opposition. **Suppression requires and demands the acquiescence of the suppressed.** To bitch about the oppressor or the exploiter is the ultimate cop-out, the self-pitying way of avoiding responsibility. It's always someone else's **fault**. Takes two to tango. The oppressor and the oppressee. Both equally necessary, or the silly little game can't be played.

The lesson of history is equally plain. Tyranny can flourish only with the passive acceptance of the tyrannized. And that, in the perspective of historical time, uneasily held balance that it represents, is **always** short-lived.

The only difference in this age of ours from the earlier ones is the addition of a frightening ingredient, the ultimate doomsday machine, our nuclear toys that, like Pandora's box **must** be opened, attractive mystery that they are.

It is an article of faith with me, this **power** that John Q. Public has. No one, no institution, no government, no giant corporation can buck John Q. I

Conspiracies thrive only in darkness

Tyranny can flourish only with passive acceptance

The difference is our nuclear toys

John Q. is really all powerful

123

think **he's** tired of the Armageddon always being held threateningly over his head. The solar conspiracy is for huge stakes, and the players are massively powerful. But the players are bureaucracies made up of John Q's, and I think **those** John Q's are tired of the old song of profits and security being the ultimate goal.

I think John Q means it

I hear John Q. saying he wants independence. I think he means it. Maybe not. The next few months and years will spell the answer irrefutably. Let's watch and see. Will the phrase **"personal independence"** take on a new meaning, one tied in the hearth and family, where, with enlightened selfishness, the head of the household begins to make the home a self-sufficient castle?

Bullheaded and stubborn

If John Q. decides that proliferation of nuclear subsystems is repugnant for the prolongation of the Doomsday threat that it represents, he'll move. Lethargically at first, no doubt. But positively. Stubbornly and immovably. That's the common man's forte. Bullheadedness in the face of unbelievable odds. That has been America's forte, as a nation of ordinary men. Too stubborn to acknowledge defeat, bullheadedly continuing regardless of cost.

Those who would try to implement the control of the sun should beware, because **this time** the game went too far. They've toyed with that principle they should not have—stealing that which every man knows in his instinctual heart to be his own...the sun.

Gratitude is expressed for the permission of International Solarthermics Corporation to reprint the four appendices of consumer information which follow.

17 Ways to Cut Your
Fuel Bill in Half

HOW
TO STAY WARM
IN AN
ENERGY CRISIS

17 WAYS TO CUT YOUR
FUEL BILL IN HALF

Nobody likes facing an energy shortage. Yet study after government study has shown we are nearly out of traditional home heating fuels. Every report seems more pessimistic than the last. Boil it all down and it means sooner or later you'll probably own a solar furnace.

In the meantime, there are plenty of ways to conserve precious fuel and cut your present heat bill. These 17 tips could save you 50%, now! And with fuel bills skyrocketing, the money you save could buy a solar furnace.

One more thing: conservation isn't as romantic as solar heating but it's just as important. Not only does it save money but the cost of the improvements increases the value of the house. You should get it back — with interest — when you sell.

17 WAYS TO CUT YOUR HEATING BILL UP TO 50%

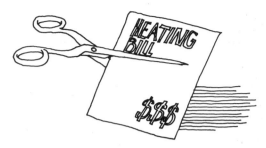

1 **Using your drapes to collect solar heat.**

Most of the sun's energy is in the form of *light* you can *see* with your eyes. When this light is absorbed by objects—like when you stand in the sunshine—it's converted to heat. Your windows let in this light—even on the north side of your home—*if your drapes are open.* Some light comes in through windows even on cloudy days.

So the first rule of energy conservation is using the windows to help supplement home heating. That means opening drapes at dawn *every day.* This will permit entry of about 100,000 Btu's per day through the windows in the typical home!

At night, however, this gain will be erased if the drapes remain open. *So you must close all drapes at night.* If you have *any* windows which don't have drapes, get them. In addition, lining your drapes adds to their insulating value and is an investment that really pays off.

2 **Cutting heat losses through windows and doors.**

Heat losses through and around windows and doors is substantial. These losses can be reduced by adding well-caulked and sealed storm windows and doors. When purchasing the storm sash, buy from a lumberyard or hardware store and install it yourself or get a friend to do so. Compare prices! You may pay a premium of several hundred dollars to have a direct sales company do the installation.

Most windows in your house are never opened in the winter. Caulk those windows thoroughly, since even well-sealed window units may permit air infiltration. Through-the-wall letter chutes should be sealed permanently and replaced by outside letter boxes.

3 If you can't afford storm windows and doors.

For less than $20 you can obtain clear plastic film from a lumberyard or hardware store to cover your windows. Tape this film on the *inside* of your windows. Tape it to the *frame* with an air space between the glass and the plastic. Don't scrimp on the tape: you want a *dead air space.* If done carefully, the plastic is invisible. And if applied on the inside, it won't discolor, crack or fade. For low budget storm doors, use plastic-covered screening. This is considerably cheaper than the tempered glass now required for storm doors. Cut to size and tack it to a screen door frame. With a bit more effort, you can tack or staple the material to light wood frames, then screw the frames onto the screen door so that they can be removed and stowed in the summertime.

4 Weatherstripping.

This is a tried and true way to reduce heat losses from around outside doors. *Even if you have storm doors* you should apply good weatherstripping. Stopping cold drafts saves energy in two ways. First, your furnace doesn't have to heat up cold outside air. Second, you feel more comfortable, so you can lower your thermostat a bit.

5 Electrical outlets.

Electrical outlets are really holes through which cold drafts can flow. Sealing them is a quick and inexpensive way to save energy. All you need is a roll of 3-inch-wide masking tape and a razor blade. First, shut off the circuit breaker and remove the wall outlet cover. Apply masking tape over the entire receptacle and make sure that the masking tape is adhered totally to the wall surrounding the receptacle box. Replace the cover and screw into place. Using the razor blade, trim the exposed masking tape away. Do a careful job and your taping is invisible. Proceed around the house taping every wall outlet and switch plate.

6 Caulking.

This is a good Saturday afternoon job. It takes some time but the energy savings are well worth the trouble. You'll need to buy or rent a caulking gun and get a couple of cases of tube caulk. *Where to caulk?* This is common sense. The goal is preventing air infiltration into your home. To do this go around inside your house on a windy day and pinpoint where air is entering. Even a small draft can be responsible for a lot of heat loss. Look at these places: the point where flashing overlaps the foundation; around the window frames; inside the house where the foundation and mudsill come together; and under the eaves. Even *under* the window sills inside your house, if you feel a draft. Caulk comes in different colors. If possible, pick one that won't require touch up.

7 Your fireplace.

This is a monster heat thief. Even with a fire going, you may be losing more heat up the chimney than the fire is adding to your home. When you have no fire, the damper makes a poor seal and losses from the house continue. There is only one *good* solution. ADD GLASS DOORS TO YOUR FIREPLACE. The fireplace will *not* be less enjoyable. In fact, with better draft control provided by small dampers under the glass doors, you will use less fuel and your fire will last longer.

8 The electric clothes dryer.

This appliance uses lots of electricity. It doesn't make sense to vent all that heat to the outside. During midwinter months, pull the exhaust hose and plug the opening to the outside. Put an old nylon stocking over the hose to catch lint and let it vent to the inside of your home. You pay for that heat—use it inside your home. This also adds needed humidity during midwinter months. CAUTION: If you have a gas dryer, have a professional serviceman check it before venting to the inside. Otherwise you may end up gassing yourself.

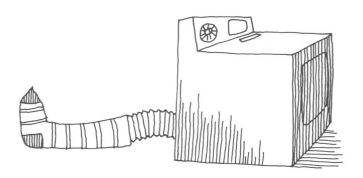

9 Your range vent fan.

Here's a real villain for stealing heat if it vents to the outside. It not only steals heat you've paid for from the range, but from the room as well. Change it so it vents through an activated charcoal filter back to the inside of the house. It will still remove odors plus give you savings on your heat bill.

10 Bathtubs.

When you take a bath, don't pull the plug when you finish! You've *paid* for the Btu's to heat the water. So let the tub water cool to room temperature (cold to the touch) before draining. Approximately 10,000 Btu's will be added to your house from a tub of water as it cools from 100 to 70°F! That's enough heat to keep your well-insulated three bedroom home warm for one hour when it is 10° above zero outside!

11 Insulation.

This is the most important point of all. You may be sure that your house is poorly insulated. Here's what you should have in the attic: 18 in. of fiberglass batt insulation or 10 in. of polystyrene foam or 6 in. of polyurethane foam. NO LESS! Batts with foil are usually the cheapest and easiest to install. Spend an afternoon and do this as soon as possible. If you have a crawl space, put 12 in. of fiberglass batt insulation between the joists. NO LESS! (This holds true whether you live in the North woods or the desert Southwest.) If you have 2 in. of batt insulation or more in the walls, it probably is not economical to add more. If you have *no* insulation in the walls, resign yourself to the expense and start shopping to get the job done for you unless you are very handy. Note: If using foam insulation, insist on a fire-spread rating of 25 or less. Second note: Beware of insulation salesmen. Insist on seeing independent laboratory tests when amazing claims are made.

12 Your forced-air furnace.

Right here you can save 15 to 30% by switching to Continuous Air Circulation (CAC). For about $15 you can get your local furnace dealer to help you, if need be. Continuous Air Circulation provides very even heating much like hot water heat. At present, your forced-air furnace blower cycles on and off, putting very high temperature air into your house at intervals. With CAC, your blower runs 24 hours per day at a much slower speed. On CAC, less heat is needed, since losses are less. Also, the heat that builds up along the ceiling — and can't be used — is pulled down and circulated along with furnace air. CAC does have a problem: like hot water heat it is slow to recover. For example, if you leave the door open for half an hour, it may take 15 to 20 minutes to regain the thermostat setting in your house. One thing to get used to: the hot air coming out of the registers is typically 72 to 78° which will feel cool on your hand.

13 Playing with your thermostat.

If you have solid masonry walls, don't. If you have typical framed house construction (even with brick veneer), it may be worth real money to fiddle with the thermostat. Here's how and when you should: if no one is at home in the daytime, turn the thermostat down to 60°F when you leave for work. Turn it back up when you return. You won't find it at all uncomfortable to do this, particularly if you opened all the drapes before leaving for work. Turn the thermostat down to 60 to 65°F when you go to bed. Experiment a bit — you'll find a level that doesn't make getting out of bed cold and miserable. You'll sleep better, too. If you take the time to fiddle, you've earned the right to set the thermostat at 72°F while watching TV, which is much more comfortable than 68°F. (If you are active, 68°F is dandy, but reading or other passive activities require a higher temperature for comfort.) If you have framed walls, it is totally a myth that you will use more energy fiddling with the thermostat. You'll save.

14 Humidifiers.

The controversy about whether a humidifier helps conserve energy or not is still not resolved. If you want one, get one. If you don't, don't. Your personal comfort should be the decider.

15 Window tint films.

Never put these on your home unless advised to do so by a registered professional engineer. They do help keep the house cooler in summer, but *raise* your heating bill in the winter. If you use them at all, they should *not* be permanent. You definitely want to remove them in winter.

16 Baseboard heaters and wall registers.

Unfortunately, radiators and registers have a way of ending up *behind* draperies. That costs you money. Devise common sense deflectors out of aluminum foil or sheeting to get the heat away from behind the curtains!

17 Shading.

It is a good idea for the south side of your house to plant trees which lose their leaves. Then they can shade the windows in summer, but not in winter. If you use awnings, try to locate them at the proper height to shade your windows in summer when the sun is high, but not in winter when the sun is low.

These 17 energy saving tips will cut your use of fuel by 50% or more. Is it worth the bother? Look at last year's heating bills and decide if a weekend's work around the house and some small and larger expenditures make sense.

What about solar furnaces? You are going to have one on your house sooner or later. When you can afford one, get it. The prices will be going up year by year with inflation. They are already in mass production in reliable, dependable form by many manufacturers. Check performance claims carefully. See the brochure entitled "Wanna Buy a Solar Furnace" in this series before you buy one.

Remember, every addition you make to your home increases equity—you'll get it back with interest when you sell. In addition, you can be comfortably warm in your house and not end up in the poor house. Not only that, you'll be making a positive contribution to making this country more energy independent.

If, as a homeowner, you're considering solar heating . . .

ADVICE TO THE HOMEOWNER OR BUILDER CONSIDERING SOLAR HEATING

You are starting at the right time! It is much easier to prepare a house for solar heating *before* you build. Following these common sense suggestions will save you *tens of thousands of dollars* — literally.

1 Locating the house on the site.

Laws regarding sunshine rights just don't exist yet. So try to locate your house on the site so that a neighbor who builds even a three-story house won't shade your solar collector. Place the house as near to the north side of the lot as zoning permits.

2 Check your house plan.

More windows on the south are obviously a benefit, but don't be afraid of windows on the north wall. They let in diffuse sunlight too, and with draperies to close at night, will not hurt you. If you can reconcile yourself to getting along without sliding glass doors, do so, because they are a substantial heat loss area. Cathedral ceilings and celestories should not be part of the plan, period. Use your imagination to find another way to get the feeling of spaciousness. You should *never* have more than an eight-foot ceiling, unless you have a second-story opening into the higher ceiling area and a way of convecting that heat away from the high ceiling area and into a living area. Fireplaces are OK, but specify glass doors on them. Range vent fans should vent through a filter back to the inside of the house. About windows: You should permanently glaze *all* windows so they cannot be opened. A small screened 4 in. x 12 in. opening near the ceiling and another near the floor which can be opened and closed will give you *better* ventilation than an open window, because natural convection is assisted. Always dual glaze.

B5

3 Insulation.

Sidewalls should contain full-width 3⅝ in. fiberglass batt insulation with foil backing. Polystyrene foam sheathing should be used *instead* of the typical impregnated board on the outside of the walls. Mudsills, plates and caps should be caulked thoroughly before the building is closed in. Note: Special attention should be given to caulking around windows and door casings. Siding joints should be caulked *during* construction. A polyethylene vapor barrier should be installed behind dry wall and should be taped to outlet and switch boxes. Knockout holes in these boxes should be caulked after wiring is in place. If the home has a crawl space, 12 in. of fiberglass batt with foil or equivalent insulation should be installed between floor joists. 18 in. of fiberglass batts with foil or equivalent should be specified for the attic. If these numbers seem large, remember that *insulation is cheap* compared to the cost of the solar heating system necessary to replace its lack.

Improve the weathertightness and insulation of your home *as you build it.* It is easier to do it now than later. Solar furnaces cost about 40 to 45 dollars per square foot of collector for the total system including installation labor. Decreasing the heat loss of the home reduces the number of square feet of collector required. Conservation and insulation are not as romantic as a solar furnace but contribute significantly to its efficiency. Keep that in mind, and you won't go wrong.

4 Forced-air heating.

Specify an adequate forced-air system be installed in the home. This may be a fueled system, but probably an electric forced-air system is a better choice in the long run. This system should be designed to handle 100% of design temperature requirements and should be operated in a CAC (Continuous Air Circulation) mode. CAC gives the evenness of hot water heat and will reduce the home's heating requirement 15 to 30%. Be sure that adequate return air ducting is provided. If a fueled, as opposed to an electric, system is utilized be sure that outside air ducting for the combustion chamber is included in the plans.

As to your choice of solar furnace, costs are skyrocketing beyond the realm of practicality if you heat with hot water. Why? The hydronic solar collector must operate at high temperatures. And as temperatures rise, collector efficiency goes down. You will need at least three times the collector area if it operates at 200°F than at 100°F. Much less expensive forced-air systems are now available that operate in the lower temperature ranges. You should investigate thoroughly before specifying any solar device.

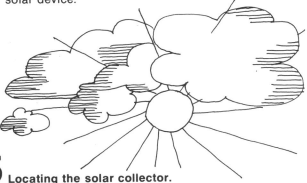

5 Locating the solar collector.

Despite the fact that many systems have roof-top collectors, you should, if at all possible, locate the collector at ground level. The 10 to 30% additional radiation obtained from reflected albedo or ground reflectance is lost if the collector is roof-mounted. Practical systems will have mirrors, as well, to reflect additional radiation into the collector, and ground-level mounting facilitates the use of these mirrors. Ideally, the collector should be oriented 10° west of due south. Common sense is important, however. For example, if a neighboring building or surrounding terrain will shade the collector in the late afternoon, orientation east of due south may be desirable.

6 Sizing the solar furnace.

Manufacturers will publish tables showing the percentage of heat requirements that can be supplied by his various models on different sizes of houses, so that you may determine the best value versus cost to obtain the amount of solar supply you desire. Note: Solar furnaces are auxiliary systems. To go from 90% solar supply to 100%, for example, might require quadrupling the size of the system, so that in effect you are buying three total systems to get that last 10%.

Reputable solar furnace manufacturers will publish average daily Btu outputs for their system in your city and will offer performance warranties as well as product warranties. If they are unwilling to do so, beware.

Now is the time to hire a registered professional engineer. Have him perform a heat loss calculation on the home to be built. He will be able to tell you the heating requirements in Btu's per hour at design temperature. Multiplying that number by the appropriate conversion factor below, you will have the Degree Day (DD) size of your home (units of Btu/DD).

CONVERSION FACTORS FOR HOURLY BTU HEAT LOSS TO DEGREE DAY SIZE

	Design Outside Temperature (°F)						
	−30	−20	−10	0	+10	+20	+30
Conversion Factor	.25	.28	.32	.37	.44	.53	.69

7 Stay away from prototypes.

Unless you are involved in the research and development business, you probably want to know how well the system will work *before* you add it into your house. For this reason, you are well-advised to purchase a total, integrated and tested *system* that includes collector, storage, heat transfer and controls. A given collector, for example, will work very differently in conjunction with two different storage systems or two different control systems. Buying *components* from different manufacturers and relying upon dubious data to hook them all together to get a working, reliable system is risky and foolhardy. Costs can really get out-of-hand when this kind of component matching is done on site. In particular, be wary of systems that utilize pebble bed heat storage. Air flow through pebble beds is complex, and changing the shape of containers for pebble beds can introduce nightmarish air flow engineering problems. Specify that only a factory-produced pebble bed container be utilized, and then integrate the *container* into your plans.

8 Do-it-yourself kits.

If these are factory-produced, you can check them out *before* you buy. Read the instructions to see that you have the skills necessary. You should *still* get a warranty, although it may include the additional words "if installed in a workmanlike manner according to instructions".

9 Building permits.

The solar furnace you install should conform to existing local building codes. If it doesn't, expect to have to appear before the county commissioners to petition for a variance.

10 Safety.

The professional engineer you hire can advise you here. Another good way to see if the proposed system is safe is to see if the manufacturer has a large product's liability insurance policy insuring him against harm to the public by his product.

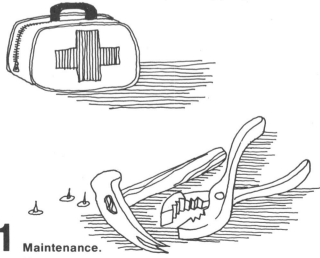

11 Maintenance.

Use some common sense to evaluate the proposed solar furnace. What kind of maintenance and repair is it going to require? How much will it cost per year to run the system? How much bother is involved? Is it totally automatic? How reliable is it?

12 Financing.

You will probably have no trouble whatsoever incorporating a solar furnace in your home mortgage. Savings and loans, commercial banks and mortgage lenders will look favorably on a solar-heated house, since high fuel payments won't interfere with your ability to make house payments! In some institutions you will even get a preferred interest rate, because they feel that conservation steps and solar heating are contributions to the local and national good.

13 Taxes.

Many states have passed legislation giving property tax breaks to homeowners with solar heating. Federal legislation is being proposed to give income tax deductions as well. It wouldn't hurt to drop a line to your Senators and Representatives to spur them on!

Advice to the
Architect Considering Solar Heat

TIPS ON DESIGNING AN EFFICIENT AND PRACTICAL SOLAR HOME

Practical solar heating is here now. With its successful introduction, a new era of residential design has dawned.

For the professional architect, solar design is both exciting and challenging. For his customer, solar heating represents a salvation from high fuel bills and a significant first step towards energy independence.

The prime precept in designing a solar residence is ENERGY CONSERVATION. Fuel costs are increasing dramatically and controlled solar heating systems are 40 to 45 dollars per square foot of collector and accompanying system installed. Obviously, the challenge is to keep heating requirements at a minimum so energy costs can be realistic.

The following pages offer 16 tips on designing an efficient and practical residence. Hopefully they will be of value to you and your staff. For the names and addresses of solar manufacturers currently producing tested solar furnaces, please feel free to write:

International Solarthermics Corporation
Box 397, Nederland, Colorado 80466

1 Do not design cathedral ceilings or celestories into the structure.

Although a feeling of spaciousness is desired, high ceilings, particularly ones with windows at the top, are tremendous heat thieves and should be avoided. Skylights should also be eliminated unless automatic insulated shutters are provided to cover them at night. Obtain feelings of spaciousness through the limited use of large windows in specific areas or the use of color or graphics. The "cozy" feeling is much better in terms of heat loss! If a cathedral ceiling must be used, recover the heat for second-story bedrooms and provide forced or natural convection.

2 Design automatic (CdS cell controlled) insulated shutters for all windows.

This is a radical design step, but one which can contribute significantly to the reduction of heat loss. Closet door framing of insulated panels motorized and controlled by photo cells to open in the daytime and close at night are not terribly difficult to integrate into the construction.

3 Have all windows permanently glazed.

Provide for screened ventilation ducts near the ceiling and floor above and below windows. The ducts must be outfitted with well-sealed insulated shutters to allow for closing in the winter. The ventilation will be superior to having half of the window itself open and will reduce the cost of windows in that sheet glass can be framed-in on site. Caution: Be sure the glazier washes the insides of the windows carefully, since they will not be accessible later. Specify either silicone or butyl caulking around glass and frame.

C5

4 Design pseudo-airlock entries.

Even with storm doors, entries are large heat loss areas. Try to compartmentalize entries by using mud rooms, formal entry halls or similar techniques to create "air-locks" around entries.

5 Windows on the north side.

You needn't steer away from adequate fenestration on the north walls of the building. A 3 ft. x 3 ft. window on the north side permits entry of about 2,000 Btu's per day of diffuse sky radiation and albedo.

6 Windows on the south side.

Don't go overboard with windows on the south side of the home. Too much glass there will create uncomfortably warm inside temperatures *even* in the dead of winter. Ten square feet of glass on a south wall will permit entry of as much as 20,000 Btu's per day of solar radiation. Typical fenestration of the past 20 years is adequate to provide nearly all the daytime-heating requirement of a well-insulated house.

7 Eaves.

Use eaves to control entry of sunlight into the building. A two to three foot eave will permit entry of winter sunshine but prevent entry of most of the summer radiation. The following table of solar zenith angles will assist in properly positioning the eaves.

SOLAR ZENITH ANGLES
(Angle between the sun's rays and the vertical)

Latitude	30°	35°	40°	45°
Summer Solstice	6.5/40.5*	11.5/40.5	16.5/41.1	21.5/42.2
Spring & Fall Equinox	30.0/52.2	35.0/54.6	40.0/57.2	45.0/60.0
Winter Solstice	53.5/68.8	58.5/72.4	63.5/76.1	68.5/79.8

*First number is the solar zenith angle at solar noon; second number is the angle at 9 a.m. and 3 p.m. solar time. For instance, on the summer solstice at 30° north latitude, the zenith angle is 6.5° at noon and 40.5° at 3 p.m.

8 Wall areas.

One challenge in designing energy conserving homes is minimizing outside wall area. Obviously, the circular home has the least outside wall area per unit area enclosed, but more conventional structures may be desired. The more square it can be, without wings, the better. Thus, the challenge becomes one of designing an aesthetically pleasing box, perhaps using roof-lines to move the eye from the regular structure.

9 Site planning and micro-environmental tempering.

Unfortunately, sunshine rights have yet to be legislated. Therefore, the structure should be located as near to the north lot line as zoning permits to prevent future shading of the collector by neighboring structures, trees, etc. Time should be spent on site, trying to anticipate changes in the future which might reduce the amount of sunshine falling on a located collector. Try to determine prevailing wind conditions. If the site is subject to prevailing northerlies, put a berm, grove of evergreens or solid privacy fence on that side of the home. Locating the garage on the "weather" side of the building to intercept prevailing winter winds is also an excellent tempering device.

10 Solar insulation standards.

Although it is easy to specify 2x6 framing on 24 in. centers to permit more wall insulation, the contractor may balk. 2x4 framing with 3⅝ in. fiberglass batt with *foil* will achieve the same results if polystyrene foam sheathing is used instead of the usual impregnated board on the outside of the structure. 18 in. fiberglass batts with foil in the attic and 12 in. between floor joists above the crawl space or an equivalent insulation should be specified. Weatherstripping and caulking should be emphasized on prints, particularly around mudsills, plates, caps, and window frames *before* the house is closed in with siding.

A polyethylene film vapor barrier should be installed behind drywall and taped to outlet boxes and switch boxes. Knock-out holes for wiring should be caulked or foamed after wiring is in place.

11 Kitchen vent fans.

These should be self-venting types with charcoal filters and should *not* be vented to the outdoors.

12 Fireplaces.

These should never be installed without simultaneous installation of glass doors to reduce heat losses up the chimney.

13 Furnace specification

Only forced air should be specified, and it should be installed to operate in the Continuous Air Circulation (CAC) mode. It makes sense to locate the furnace room close to the proposed location of the solar furnace. Registers, if located under windows, should be specified with deflectors or set far enough from the wall not to blow behind draperies. Electrical forced-air furnaces or in-line electrical duct heaters are recommended for use with solar furnaces. If a fueled forced-air system is specified, provide for and specify outside ducting to the combustion chamber. Check to see that return air registers are properly located and adequate in size. You will have to educate your customer about CAC mode heating. The blower runs 24 hours a day at a much slower speed, and supply air is much cooler (72 to 78°F typically) in this mode (20 to 27°F cooler than body temperature). CAC is very much like hot water heat, i.e. very even, but like hot water heat the response time is somewhat slower. CAC will save 15 to 30% on heating bills.

14 Integrating the solar furnace into the structure.

Only as a last resort should the collector be located on the roof since radiation inputs will be less there than at ground level. Reflectance from the ground and surroundings will generally not impinge on a roof-mounted collector. Typically the collector size should not be larger than 160 square feet for a well-designed and insulated home, probably much less. The solar furnace may be integrated into the south wall of the structure, with the collector appearing to be a bay window. It may also be totally separated from the structure, integrated into the back of a garage or carport with tool storage provided, or used as a design element to create a wing effect. This design element can then be harmonized with other aspects of the structure.

15 Note about pebble bed storage systems.

Many solar furnaces specify a pebble bed storage battery. It is risky, if a fine pebble bed storage system is being utilized, to use other sizes and shapes of containers than those specified by or produced in the factory. Air flow through a fine pebble bed is tricky, and trying to engineer these flows on site is a prototype game which may inadvertantly involve you in a time consuming and costly R&D business. If pebble size specified by the manufacturer is greater than 1½ in. you should be wary, since although larger rocks permit better air flow, the heat transfer surface area is significantly reduced and more storage mass than necessary will be required. An efficiently engineered pebble bed will employ stratification and counter-flow heat exchange principles. If it doesn't, beware.

16 Sizing the solar heating system.

It is recommended that you retain the services of a registered professional engineer with HVAC and solar experience. He can provide you with a variety of size versus price versus percentage solar supply choices, so that you may make an educated decision. For the protection of the customer, insist on a structural warranty *and* performance warranty. Typically the performance warranty should read as follows:

   a. **Collection Capability**—Cloudless day heat transfer to storage of ... Btu's/day on ... (dates) ...with an average daily shortwave radiation of... Btu's/ft² on a horizontal surface and an initial storage temperature of ... °F.

   b. **Storage Capability**—Delivery of useful heat of ... % of the energy stored before a 72-hour period of stagnation during which the average air temperature surrounding the storage container is ... °F.

Insist on the Btu output rating, so that you may tell the customer how the unit will perform *before* it is installed.

Most of these recommendations are common-sense variations on standard building practices, although a few radical ideas have been included. Solar heating systems are expensive, and therefore, the house should be designed for minimum heat loss to bring the size of the solar furnace down as much as possible. Good, practical design and engineering at the conceptual start of a building can reduce its heating requirements by as much as 70% and will minimize cooling loads as well!

Myths and Facts
About Solar Heating

Myths and Facts
About Solar Heating:

- MYTH: "Practical solar heating is 10 years away."

- MYTH: "100% solar heating is impossible."

- MYTH: "The solar collector must go on the roof."

- MYTH: "You need a million Btu's per day to heat a three-bedroom house."

Solar Heating...a New Science

Fuel shortages and rising costs have rekindled man's interest in harnessing the sun. In the past few years, millions of dollars have been spent to learn the properties and potentials of earth's nearest star. Unfortunately, scanty data from scattered research projects have often been quoted out of context. Consequently, partial truths and old wives tales based on preliminary research from another era have become the myths of today.

The fact is that solar heating is here now. It can work. It does work.

This booklet of widely quoted myths and their accompanying facts will help you weed truth from fiction and put to rest the mythology of another era. Hopefully, it will prove to you that harnessing the sun is now much more than a dream.

MYTH: "Practical solar heating is ten years away."

FACT: This was true, until International Solarthermics Corporation, a solar engineering firm, devised a system compatible to existing homes and new housing developments. Manufacturers are now producing this practical, backyard solar furnace that can provide up to 90% of the home heating requirements.

MYTH: "100% solar heating is impossible."

FACT: With present technology, 100% solar heating is possible. It is not, however, economically practical. Going from 90% efficiency to a unit capable of total home heating might triple or quadruple the cost of the furnace. It is very expensive to provide the necessary reserves needed for unusual weather, long periods of cloudiness or extreme cold. For that reason most solar furnaces are auxiliary units and designed to complement, not replace present heating systems.

D4

MYTH: *"You need a solar collector with one-half the square footage of the building it will heat."*

FACT: The size of the collector needed is dependent on three basic factors: 1) the efficiency of the collector, 2) the heat storage medium used and 3) the heat loss of the building.

For example, if the collector can absorb only 200 Btu/ft² day, it probably will have to be half the size of the building. However, if it can collect 2000 Btu/ft² day, it can be ten times smaller. The same is true of storage. Rapid cooling of the battery obviously requires a much larger solar collector to compensate for heat loss. The heating needs of the building are also a factor when engineering the solar collector.

MYTH: *"Hydronic (water) systems are the best way of collecting and storing solar heat."*

FACT: The hydronic solar heating system is certainly the most popular and the one most often seen being tested. It uses water as a heat storage medium and nearly all tests to date have been conducted using the hydronic systems. While water storage is the most popular, it is not necessarily the best or most efficient for the following reasons:

1. Water is a poor storage medium for solar heating applications despite its widespread use. A typical, insulated water storage system will hold heat at a usable level no more than 18 hours. After that the water is generally too cool to continue heating the building. On the other hand, insulated rock will hold heat at a usable level for 30 days or more.

2. The high temperatures required for solar heating tend to precipitate heavy scale build ups from the water circulating in pipes and storage tanks. Examine a five year old hot water heater and you'll discover an inch or more of scale inside. This scale in a hydronic solar system can cause exhorbitant maintenance costs.

3. Hydronic systems can require large quantities of expensive commercial anti-freeze—some systems use as much as 2500 gallons at a cost of approximately $15,000. That initial expense

D5

is ridiculously high particularly when alternative storage mediums are thousands of times cheaper.

4. Most American homes are heated with hot air. With the hydronic solar furnace, the air must be heated during a second transfer step. This decreases efficiency and increases cost.

MYTH: *"All flat plate collectors generate about the same amount of heat per square foot."*

FACT: A collector's capability to absorb and give off usable heat is dependent on its configuration, the effectiveness of its heat transfer system and the degree to which losses from the system are hindered. In the past, flat plate collectors have operated in the general range of 50-200 Btu/ft² day. International Solarthermics Corporation engineered their collector to work more efficiently—up to 2000 Btu/ft² day. Their Thermal Transceiver™ unit uses vertical vanes, significant loss reductions from the collector and an extremely efficient heat transfer method to generate these increases.

MYTH: *"The solar collector must go on the roof."*

FACT: If the collector is large—say 1000 square feet or more—the homeowner has little choice but to put it on the roof. The location of the collector then becomes a function of available space rather than efficiency. Actually, roof-top collection is often impractical. The weight of the collector requires considerable structural reinforcement and often the pitch and orientation of the roof is incorrect.

With existing homes, where the roof is simply not engineered for the collector, a ground level unit is often the best solution. Besides, installation and construction costs are much less.

SOLAR HEATING IS A LOT OF HOT AIR!

MYTH: *"The only way to test a solar furnace is put it on a house and watch it for many years."*

FACT: Such methods of "testing" raise far more questions than are answered: how mild was the winter, did the inhabitants get more energy conscious, did they maintain the house at a lower temperature,

what do the realized fuel savings really mean? This method of testing makes every installation a prototype. And all it determines is if the house stayed warm. Proper testing, under carefully controlled laboratory conditions, can do much more. Such tests evaluate the total demand capability of a total system. It predicts efficiency under average conditions on any home with average people living there by geographical and climatological areas.

MYTH: *"Computers show that no more than 100-400 Btu/ft² day collection is possible with a flat-plate collector."*

FACT: A computer's answers are no better than the facts and premises it uses. Original experiments and data did show that early collectors could do no better than 400 Btu/ft² day. However, state-of-the-art improvements are now available. The unit developed by International Solarthermics Corporation is capable of absorbing up to 2000 Btu/ft² day.

MYTH: *"You need a million Btu's per day to heat a three-bedroom house."*

FACT: Probably less than half that amount. Home heating requirements vary with locality, severity of the winter and efficiency of the design and insulation of the house. For example, here are the requirements per day for a well-insulated 4000 square foot ranch-style home (2000 sq. ft. main floor, 2000 sq. ft. basement) for an average day in January. Note, the requirements are much less than a "million Btu's".

	Btu / day
Los Angeles	90,000
Denver	273,900
Chicago	292,500
Minneapolis	394,600
New York City	238,550

(Based on a 7500 Degree Day House and climatological data from 1931-1960.)

MYTH: *"Solar furnaces will get cheaper, so I'll wait."*

FACT: Unfortunately, probably not true. The cost of raw materials and fabricated materials goes up almost daily. Waiting for the price to go down is somewhat analogous to a man looking at a Model A Ford at $800.00 and saying "I'll wait till they get cheaper."

MYTH: *"If a system costs $4000.00, it will take twenty years to amortize the cost in fuel savings."*

FACT: Any home improvement is an investment in real estate and is *not* amortized. A new bathroom in your unfinished basement, for example, is not amortized in  flushes per month. Whatever you paid for the addition you will get back with interest, when you sell the home. Even if you don't sell the home, the additional equity is there.

Thus, with the investment in solar heating as a home improvement, you add to your *home equity.* Fuel savings are a bonus dividend on that investment.

MYTH: *"An installed price of $4000.00 is too much for a solar heating system."*

FACT: Perhaps. But alternative heating forms are much more expensive. Coal stokers, for example, are far too expensive for the average home owner. As for other forms of solar heating, most are experimental and priced well over $20,000.

MYTH: "We aren't running out of fuel—it's just an oil company conspiracy to drive up profits."

FACT: We are running out of fossil fuels. It took billions of years to create coal, gas and oil and and they can't last forever. Study the Federal Power Commission graph below and decide for yourself if we've got enough natural gas.

Consider also the other uses we make of petrochemicals for plastics, man-made fabrics and other uses. Then decide which is a better use of petroleum: to burn it up or create products with it.

UNITED STATES NATURAL GAS SUPPLY—DEMAND BALANCE
(Contiguous 48 States)

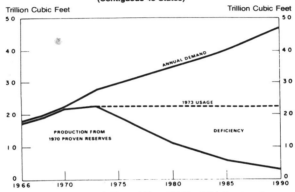

SOURCE: Adapted from Figure 1, Page 3 National Gas Supply and Demand 1971-1990

FPC Bureau of Natural Gas, Washington, D.C. February 1972

MYTH: "The government has spent millions of dollars in research. No private company using private funds could accomplish what all the scientists have failed to do."

FACT: Many of the major scientific discoveries of the world have occurred through private research. The Wright Brothers, Henry Ford, Edison and Alexander Bell all made significant contributions as "backyard tinkerers." The discovery of vulcanized rubber, Scotch Tape and the rotary engine came from the dedicated research of a handful of privately funded scientists.

Today, International Solarthermics Corporation has made a major technological breakthrough in solar engineering years before the experts predicted it could happen. Why? Perhaps private capital is spent more prudently than

D9

government funds. Perhaps it was a lucky break, much like the discovery of penicillin. As with many such discoveries, the scientific principles are actually quite simple. Perhaps that's why it has been overlooked by the nearly 1000 federally financed projects now going on.

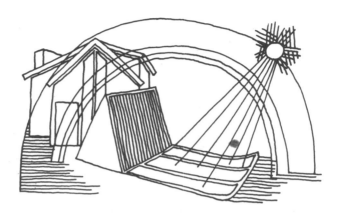

MYTH: "Every house in America is ready now for solar heating."

FACT: Totally untrue! Old-fashioned insulation standards simply don't apply to the new technology of solar heating applications. To prepare a house for solar heating, one should add enough insulation to the attic to bring it to 14 inches. Solar-heated homes should also have storm windows, weather stripping and common-sense sealing of all openings. Proper insulation never costs as much as the added expense of doubling or tripling the size of the solar furnace needed to get the same effective heating. Besides, even if solar heating is not added, the improved insulation will help preserve our diminishing supply of fossil fuels.